MAPS AND MEMORY
IN EARLY MODERN ENGLAND

Early Modern Cultural Studies

Ivo Kamps, Series Editor

Published by Palgrave:

MAPS AND MEMORY
IN EARLY MODERN ENGLAND

A SENSE OF PLACE

Rhonda Lemke Sanford

palgrave

First published 2002 by PALGRAVE™
175 Fifth Avenue, New York, N.Y. 10010 and
Houndmills, Basingstoke, Hampshire, England RG21 6XS.
Companies and representatives throughout the world.

PALGRAVE is the new global publishing imprint of St. Martin's
Press LLC Scholarly and Reference Division and Palgrave Publishers
Ltd. (formerly Macmillan Press Ltd.).

ISBN 0-312-29455-7

Library of Congress Cataloging-in-Publication Data
Sanford, Rhonda Lemke.
Maps and memory in early modern England : a sense of place / by
Rhonda Lemke Sanford.
    p. cm.
  Includes bibliographical references and index.
  ISBN 0–312–29455–7
  1. English literature—Early modern, 1500–1700—History and
criticism. 2. Place (Philosophy) in literature. 3. Cartography—
England—History—16th century. 4. Cartography—England—
History—17th century. 5. England—Maps—History.
6. Geography in literature. 7. Memory in literature. 8. Maps in
literature. I. Title.

PR428.P56 S26   2002
820.9'32—dc21

                                                          2001036572

Design by Letra Libre, Inc.

First edition: April 2002
10 9 8 7 6 5 4 3 2 1

Printed in the United States of America.

For my parents,
Darrell D. Lemke
and Frieda Lemke

Contents

LIST OF ILLUSTRATIONS

ACKNOWLEDGMENTS

It gives me great pleasure to thank the many people who have had a hand in shaping *Maps and Memory* and who have supported me in various ways during the writing and revising process. I would like to thank several wonderfully gifted early modern teachers and mentors: Katherine Eggert got me started thinking about both *Cymbeline* and the subject of traveler's tales that eventually led me to maps and much more. Margaret Ferguson has helped to shape my thinking and writing in more ways than I could list. I thank them for directing my dissertation, which is at the heart of the current book, and for being just the kind of generous scholars I'd like to be. I cannot imagine better mentors and role models than these. R L Widmann sparked my interest in early modern literature and especially in the country house poem; Richard Halpern sustained my interest, and challenged me at various steps through the completion of this project. At the University of Colorado, I would also like to acknowledge the support of Clare Farago, Jeffrey Robinson, Elizabeth Robertson, John Stevenson, Charlotte Sussmann, and Eric White.

I would like to thank the University of Colorado for the following fellowships in support of my research: The J. D. A. Ogilvy Fellowship for travel to Britain, The George F. Reynolds Fellowship, The William B. Markward Fellowship in Renaissance Literature, and the Emerson and Lowe Humanities Fellowship.

For help with questions on Spenser and comments on early versions of chapter 2, I thank Patrick Cheney, Susan Frye, and Richard Helgerson. I would like to thank Patrick Cheney for including me on a panel for the 1996 MLA and Richard Helgerson for commenting on a very early version of my reading of the river marriage episode. Thanks to Virginia Mason Vaughan and Alden Vaughan for the SAA session "Playing Across the Globe: The Geography of English Renaissance Drama" in 1996, and to Virginia Mason Vaughan and John Gillies for many good ideas about earlier versions of the *Cymbeline* material that appeared in their book, *Playing the Globe, Genre and Geography in English Renaissance Drama*. Thanks to Fairleigh Dickinson University Press for permission to reprint the revised version of that essay. I thank conference organizers Andrew Gordon and Bernhard Klein of the Paper Landscapes conference in London where I presented early work on Isabella Whitney and learned so much from all of the papers and participants. I would also like to thank seminar leaders, Mary Bly and Katharine Eisaman Maus, and seminar participants in the SAA Seminar "Mapping the Geographical, Theatrical Margins of London," which was very helpful on chapter 5.

Librarians have provided laudable help not only in research, but also in the final scramble with photos and references. I am indebted to Peter Barber, Tony Campbell, and Brett Dolman, and Lora Afric of the British Library; John Fisher and others at the Guildhall Library; and the staffs at the Bodlein Library, Cambridge University Library, the Museum of London, the Folger Shakespeare Library, the Burghley House Trust, and the Kunsthistorishes Museum in Vienna for permission to use the Vermeer painting that graces the cover.

I owe a debt of gratitude to the members of the online discussion group in cartographic history ("Maphist"), who have given me invaluable information, even when I'm just lurking in on the discussion, but especially when I have needed some special assistance. I want to thank Jody Giberti for helping me to get permission, and Michael and Claudia Franks for granting permission to use the lyrics from Michael Franks's delightful song, "Popsicle Toes."

Jean Howard has supported and encouraged me along the way, and I thank her for that support. Many thanks to Mac Dowdy, Kathleen McCormack, and Elizabeth White, who made a summer course at the University of Cambridge on the English country house a true pleasure.

I would like to acknowledge Ivo Kamps for his invaluable guidance, Kristi Long at Palgrave for her interest in the project, members of the editorial and production staff at Palgrave and the anonymous reader whose suggestions improved the manuscript immensely.

Friends have been a great support all through this process. I would like to thank Gloria Eastman, Jordan Landry, and Donna Long, and many friends at CU, including Sara Morrison, Teresa Nugent, Laura Wilson, Deborah Uman, Pat Kelly, Julie Radliff, and Robin and Bryan Bott, all of whom have read early drafts of this and other work of mine and for whose ideas I am always grateful. I thank my friends and colleagues at Fairmont State College, especially my chair, Martin Bond, for cheer and encouragement. I thank Steve for his unflagging support and his unflappable spirit through this and so much more, and for reading the manuscript far too many times. And finally, I would like to thank my parents, to whom *Maps and Memory* is dedicated, for giving me my own sense of place and sense of direction and for lifting me up in so many ways; I can simply never thank them enough.

Selections from Mary Carruthers's *The Book of Memory* and John Gillies's *Shakespeare and the Geography of Difference* are reprinted with the permission of Cambridge University Press.

TEXTUAL NOTE

For quotations of early modern texts, I have retained original spelling except for long s's, and I have expanded contractions made with supralineal marks over vowels.

I have included signature notations for older texts with superscripted notations for the recto (^{r}) and verso (^{v}) side of the pages. Page numbers have been provided where the edition is available in paginated facsimile.

NOTE ABOUT INTERNET SITES

I have included Internet websites as additional sources for some of the maps and images I discuss here. I have favored official and institutional websites over unofficial and personal websites. I would add the caveat that the Internet is an ever evolving entity, so that some of these web addresses as well as the content may change over time.

Series Editor's Foreword

Once there was a time when each field of knowledge in the liberal arts had its own mode of inquiry. A historian's labors were governed by an entirely different set of rules than those of the anthropologist or the literary critic. But this is less true now than it has ever been. This new series begins with the assumption that, as we enter the twenty-first century, literary criticism, literary theory, historiography, and cultural studies have become so interwoven that we can now think of them as an eclectic and only loosely unified (but still recognizable) approach to formerly distinct fields of inquiry such as literature, society, history, and culture. This series furthermore presumes that the early modern period was witness to an incipient process of transculturation through exploration, mercantilism, colonization, and migration that set into motion a process of globalization that is still with us today. The purpose of this series is to bring together this eclectic approach, which freely and unapologetically crosses disciplinary, theoretical, and political boundaries, with early modern texts and artifacts that bear the traces of transculturation and globalization.

This process can be studied locally and internationally, and this new series is dedicated to both. It is just as concerned with the analysis of colonial encounters and native representations of those encounters as it is with representations of the other in Shakespeare, the cultural impact of the presence of strangers/foreigners in London, or the consequences of farmers' migration to that same city. This series is as interested in documenting cultural exchanges between British, Portuguese, Spanish, or Dutch colonizers and native peoples as it is in telling the stories of returning English soldiers who served in foreign armies on the continent of Europe in the late sixteenth century.

At times it must have been immensely difficult for the English to locate themselves in this rapidly changing and seemingly ever expanding world. One might say that rediscovery of antiquity by nonclerical classes, the exploration of the Indies and Americas and its cultures, and the break with the Church of Rome—to name just some obvious factors—necessitated a new definition of the English nation and its people. One obvious but highly complex response to this need was a concerted but diverse effort to *map* the world, literally to commit to paper through pictorial representations where the different parts of the world were and how they were related to one another by natural and artificial borders. Another, perhaps less obvious, but no less complex response can be found in literary efforts at geographical representation, ranging from John Stow's antiquarian *Survey of London* (1598) to Shakespeare's poetic rendering of England as a "sceptred isle" and a "demi-paradise," or Ben Jonson's epigram in favor of a proposed union between England and Scotland that pictures the world as "the temple," "the priest the king, / The spoused paire two realmes, the sea the ring."

Rhonda Lemke Sanford's *Maps and Memory in Early Modern England: A Sense of Place* is a wide-ranging yet finely focused study of how the English tried to locate themselves and their nation through the creation of both maps and literary representations of geographical space. Sanford is particularly interested in the transformation of geographical *space* into a sense of *place,* and is keenly aware that the maps and literary representations that are part of this process of transformation are much more likely to *produce* than *re*produce reality. In representing place, what is positioned at the center, what at the margins—and why? What techniques and principles are called on to distinguish sacred places from profane ones, and how are shifts between the two facilitated? Any depiction of place, Sanford argues, be it visual or verbal, is foremost a metaphor, and must to be analyzed as such. Hence, Sanford offers an understanding of early modern ideologies of place derived from an analysis of literary representations of place (for instance, Jonson's poetic "survey" of Penshurst) as well as of the surveying and mapping techniques of the period that brings into sharp focus the kind of representational decisions that produced pictorial and figural depictions of *place.*

Sanford draws on Shakespeare's *Cymbeline,* Spenser's *Faerie Queene,* and Isabella Whitney's and Ben Jonson's poetry, as well as various prose tracts on map making and surveying, to argue that the metaphoric models of place (as well as particular places) created by these texts become inscribed in memory and shape how people conceptualize and experience their world. Sanford, however, also reverses this premise about the directional relationship between mapping and memory, and considers several instances in which the mapping of the land rests on memorial reconstruction (sometimes of the disenfranchised), suggesting a perhaps surprising yet crucial influence "from below." Ultimately, *Maps and Memory in Early*

Modern England delivers a compelling description of the wonderfully complex interplay between a growing desire for scientific precision and the utterly subjective nature of representation.

Ivo Kamps, series editor

CHAPTER ONE

MAPS AND THEIR REPRESENTATIONS IN EARLY MODERN ENGLISH LITERATURE

Saving your tale Syr, we poore Country-men doe not thinke it good to have our Lands plotted out, and me thinks indeede it is to very small purpose: for is not the field it selfe a goodly Map for the Lord to looke upon, better then a painted paper? And what is he the better to see it laid out in colours? He can adde nothing to his Land, nor diminish ours: and therefore that labour above all may be saved, in mine opinion.

(John Norden, The Surveiors Dialogue, *1618*)

... In that Empire, the Art of Cartography reached such Perfection that the map of one Province alone took up the whole of a City, and the map of the empire, the whole of a Province. In time, those Unconscionable Maps did not satisfy and the Colleges of Cartographers set up a Map of the Empire which had the size of the Empire itself and coincided with it point by point. Less Addicted to the Study of Cartography, Succeeding Generations understood that this Widespread Map was Useless and not without Impiety they abandoned it to the Inclemencies of the Sun and of the Winters. In the deserts of the West some mangled Ruins of the Map lasted on, inhabited by Animals and Beggars; in the whole Country there are no other relics of the Disciplines of Geography.

—*J. A. Suárez Miranda:* Viajes de Varones Prudentes,
Book Four, Chapter XLV, Lérida, *1658.*
(Jorge Luis Borges, "On Rigor in Science," Dreamtigers, *1964*
Translators Mildred Boyer and Harold Morland)

The example of the map that covers the entirety of its subject, described by "Suárez Miranda," has become an emblem of the limits of representation as well as the need thereof. Jorge Borges makes this map an example of the "rigours of science," Jean Baudrillard alludes to it in the opening of *Simulations,* and Umberto Eco uses it to meditate on its own impossibility through *reductio ad absurdum* in "On the Impossibility of Drawing a Map of the Empire on a Scale of 1 to 1."[1] If the map *can't* "cover" its entire subject area and show every detail of the original, decisions must be made as to what level of detail *can* be or *will be* depicted. What features will be highlighted, summarized, or left out entirely? These are but a few of the many decisions regarding representation with which makers of maps are faced, since maps can never merely *reproduce* reality. Even the enormous Ordnance Survey maps of England required that decisions regarding representation be made.

A complete dismissal of the value of representation is voiced by John Norden's Farmer above who believes that no such decisions need be made if the land *itself* is its own map.[2] But Norden's Surveyor and his readers already know that this cannot be the case, and in the end, as these things usually go, the Farmer too is disabused of his naïve beliefs as he becomes an enthusiastic supporter of the surveyor and allows the surveying project to move forward. It must be that somewhere between the map as large as its own subject area and the subject area as its own map (in other words, no map at all) lies a middle ground; this study addresses the representational decisions that must be made in achieving that middle ground in both pictorial and figural depictions of place, that is, in maps and other representations of place in early modern England.

Discussion of "mapping" is complicated by the fact that maps are used as both "a metaphor for knowledge and also as a major means of knowledge representation in a wide array of cultures."[3] Not surprisingly, various definitions of the word "map" have been advanced. J. L. Lagrange, the French mathematician, proposed in 1770 that "A geographical map is a plane figure representing the surface of the earth, or a part of it," and Leo Bagrow adopts this definition in his *History of Cartography* in 1963 (22). The British Cartographic Society clarified the term at its founding in 1964, first defining "cartography" as "the art, science and technology of making maps, together with their study as scientific documents and works of art"; this definition went on to say that "in this context maps may be regarded as including all types of maps, plans, charts and sections, three-dimensional models

and globes, representing the earth or any heavenly body at any scale"
(qtd. in Harley and Woodward, I: xv). J. B. Harley and David Woodward formulate an entirely new definition at the outset of their monumental *History of Cartography* (1987): "Maps are graphic
representations that facilitate a spatial understanding of things, concepts, conditions, processes, or events in the human world" (xvi).
Their concern in the *History* with how maps "store, communicate,
and promote spatial understanding" allows for a discussion of how
maps are also created in the literary realm.[4] Extending the definition
to that realm, John Gillies concludes that "in short, a map is any kind
of model of any spatial image" (54). Gillies goes on to say that

> If virtually any kind of spatial image is a map, theoretically regardless
> of its material form, then the field of cartography is not so much ex-
> panded as exploded. Modern cartography addresses this problem by
> posing the *reality* of the map in terms of process rather than product;
> in terms of a semiological (or signifying) activity rather than an inert
> artefact. (54)

The "explosion" to which Gillies refers reverberates throughout the
present study.

Besides pictorial or graphic representations, I want to suggest in
the chapters that follow that the cognitive or mental map is another
important underpinning of the literature of place. Peter Barber pos-
tulates that early modern folks were developing the ability to think
cartographically—that a kind of map consciousness was emerging
that allowed these folks to take visual possession of their topographic
surroundings. When we consider the map in terms of the process by
which it signifies meaning, the phenomenon of the mental map, or
of thinking cartographically, must also be considered. Various theories in the field of cognitive psychology suggest that cognitive maps
can be assembled from both verbal and visual knowledge.[5] Robert
Lloyd concludes, in fact, that the "only difference between the internal representations generated by a verbal description of a map and
ones generated by a visual experience would be in the amount of detail" (540). In the early twenty-first century, and in a culture where
most of us have grown up with an abundance of maps, it may be hard
to imagine not having a sense of map consciousness, or to project
ourselves back to a time when maps were not commonplace items.

Tom Conley has recently suggested that it is not merely the cartographic and scientific developments toward "transparency or accuracy
from confusion and doubt about the size and shape of the world" that

fuels our fascination with cartography of this time period (7); instead, he posits four concepts from psychoanalysis that are of use in understanding the allure of these maps. The first is the individual's relation to the unknown; that is, an individual's impulse, not unlike that which inspired the geographic discoveries of the time, that represents "a subject's ever renewed and renewing encounter with the mysteries that foster a desire to seek a sense of identity" (8). Second is the perspectival object, or the "positioning and mapping of the self in and about the world in its ongoing construction of psychogenesis," designating "a series of junctures between a viewer and what he or she sees, projects, fantasizes, and remembers, but what also always eludes containment" (13). Third are pictograms, or "fragments of visible writing in the field of memory" (17) that "[conflate] language and image" (18) and help the subject to "create an imaginary world of impressions that tie his or her body to a *mobility* of space and place" (20). And fourth, the signature, an attempt to make a personal connection "by affixing his or her proper name to signs of works, times, and places" (20). In short, understanding our relationship to the world and the word is what is at stake here.

CONVENTIONS OF MAPPING: CENTERS, PERIPHERIES, AND ORIENTATIONS

If maps are to work at all as representations, they work by means of shared understanding, assumptions, and conventions. Centers of maps are but one of these conventions. The periphery of the map, its margins, and ancillary materials are also significant to this study. Finally, orientation is an important underpinning to our understanding of space, now taken for granted, but not always so hegemonic.

Essential to my thesis is a basic understanding of the manner in which the most rudimentary maps privilege the center. Ancient Chinese diagrammatic representations of the universe consisted of a series of concentric rectangles and five directions: north, south, east, west, and middle, each of which had its own quality. The center rectangle represented the imperial palace, the next rectangle was the imperial domains, then the lands of the tributary nobles, the zone of pacification ("where border peoples are adjusting to Chinese customs"), and the land of friendly barbarians, and finally the lands of savages "who have no culture at all" (Edgerton 24). Ancient Christians placed Jerusalem at the center of their world and in the center of their maps and put east at the top, since it is the location of paradise.[6]

In a type of map that has come to be known as the "T-O" map, one of the most rudimentary of world pictures, the "circle of the earth" in Isaiah 40:22 is preserved (Figure 1.1[7]). The ocean (the "O") circles the entire land mass; inside this "O" the three known continents are separated by other major bodies of water that form a "T." Since east is at the top, the top half of the circle is Asia, separated from Africa by the Nile and from Europe by the Don—these two rivers form the top of the T, and the Mediterranean forms the upright of the T between Europe and Africa.[8] The three continents are affiliated with the three sons of Noah: Shem (Asia), Ham (Africa), and Japheth (Europe).[9] Medieval Christian maps, or *mappaemundi*, are indebted to the T-O for their underlying structure; both the T and the O are still visible, although a great deal of geographic and decorative detail has been added; the Hereford *Mappa Mundi* presents an excellent example of this addition of detail (Figure 1.2[10]). East continues to be at the top, where *mappaemundi* frequently locate paradise; and the centrality of Jerusalem as the omphalos of the world is made corporeal by representing Christ's nativity, Christ's body, and the church as central to any true understanding of the world.[11] The Psalter Map of 1250 features Christ looming above the medieval world with the center of the map, Jerusalem, positioned where Christ's navel would be if we infer his body beneath the map (Figure 1.3[12]). The implied relationship between Christ's navel and Jerusalem is made manifest in the Ebstorf Map (ca. 1235), which superimposes the medieval map *onto* Christ's body: his head is at the top, his outstretched and stigmatized hands are at the side margins, and his feet are at the bottom.[13] Samuel Edgerton tells us:

> Christ's feet are at Gibralter, the bottom and the western exit from the Mediterranean. Here the legendary Hercules is supposed to have erected two pillars with the inscription, *Non plus ultra,* "There is nothing beyond." The Ebstorf map thus represented both physically and metaphysically the *Corpus Domini*—microcosm and macrocosm united. Jerusalem is the Savior's sacred umbilicus. The good Christian viewer of this *mappamundi* was reminded of his Christian duty by the circular Eucharistic form of the diagram. He could thus never, as a God-fearing Christian, forget Jerusalem, the very Blood and Body of Christ.
>
> Nor could the good Christian think seriously of sailing away through the Pillars of Hercules, because that too would mean abandoning the Body of Christ. (29)

An analogous, though secular, figuration of man as microcosm is described by Vitrivius in his *De architectura,* written during the reign

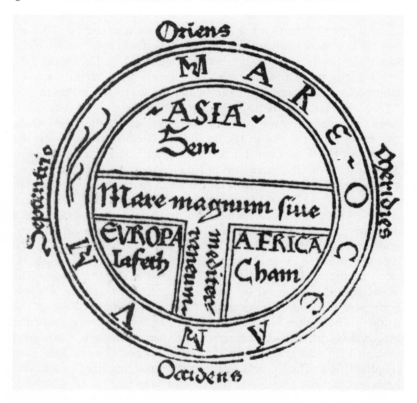

Figure 1.1 T-O map, St. Isidore of Seville, 9th Century (By permission of the British Library; shelfmark Egerton 2835)

The T-O map demonstrates a basic understanding of the tripartite world after the flood and is the basic model for medieval *mappaemundi*. In the T-O map, the ocean (the "O") circles the entire land mass; inside this "O" the three known continents are separated by other major bodies of water that form a "T." The T-O map in its most basic conception preserved the "circle of the earth" in Isaiah 40:22. The schema is oriented with East (oriens) at the top; the top half of the circle is Asia, separated from Africa by the Nile and from Europe by the Don—these two rivers form the top of the T; the Mediterranean forms the upright of the T. The name of one of Noah's sons appears under the name of each continent (Sem, Cham, and Jafeth, or Shem (Asia), Ham (Africa), and Japheth (Europe)). Medieval maps, or *mappaemundi,* are indebted to the T-O for their underlying structure.

of Augustus Caesar; Vitrivius observed that a "circle inscribed around a man with arms and legs outstretched would have its center at his navel." This model accords the navel "as point of entry for nourishment of the infant in its mother's womb . . . cosmic significance as center of the form that circumscribed not only man but the whole

Figure 1.2 Hereford *Mappa Mundi,* ca. 1300 (The Dean and Chapter of Hereford Cathedral and the Hereford *Mappa Mundi* Trust)

This map derived from the T-O map adds many detail of known geography, along with some fanciful ones. East is still at the top, Jerusalem at the center; there are still only three known continents separated by the large body of water in the middle. In the upper right hand corner is the Red Sea. In the lower right hand corner, at the edge of Africa are the monstrous races reported in traveler's tales. England is in the lower left hand corner, shown as a separate island from Scotland, with Ireland directly underneath.

universe" (Edgerton 12). In fact, there was a monument in the Forum in Vitrivius's time known as *umbilicus mundi*—the navel of the world. Tom Conley includes a lengthy discussion of the navel as an important psychoanalytic underpinning of the cartographic impulse. Using the

Figure 1.3 Psalter Map, ca. 1250 (By permission of the British Library; shelfmark ADD. 2861 f.9)

In this image Christ looms over a tripartite world centered at Jerusalem and oriented with east at top, where this map locates paradise. The monstrous races are again present at the margin of Africa (lower right-hand corner), and dragons lurk beneath the map itself.

work of Guy Rosalto (in turn based in Freudian psychoanalysis), Conley suggests that the navel is "construed to be a site where the relation of the unknown has its first noticeable, physical trace"—thus, it is a "site of a ruptured attachment to the world"; further, as the "point of separation from the mother," the navel "also becomes a site where subjects define their lives through the fantasy of losing an appendage"; finally the navel represents "a slit or an opening that is absolutely necessary for a subject to gain a sense of time and place"—that represents the desire to be "at one with his or her (self-made) 'map,' which is the world itself, being at one with the local, national, global, and cosmic space in which he or she visualizes an origin associated with a site of birth" (9–10).[14]

Leonardo's drawing of *Man in a Circle and a Square* was supposed to be an illustration of Vitrivius's text. Leonardo's drawing, however, adds a square to Vitrivius's circle and actually makes the genitals occupy the central position.[15] If this drawing can be taken as a secular descendent of the *mappamundi,* then the center here has moved from navel to genitals, a topic to which I will return in later chapters.[16] This example represents well a commonplace of the early modern period as a time of shifting centers. Besides a shift from the sacred to the secular in Renaissance humanism, the transition in science and understanding from a Ptolemaic geocentric universe to a Copernican heliocentric universe further called centers into question. Finally, the "discovery" of the "New World" and its inclusion on maps shifted the center of the map of the world from Jerusalem to the middle of the Atlantic Ocean. According to Gillies this repositioned center made the New World more enticing, as we shall see in chapter 2. Centers of maps are, in fact, frequently subject to change depending not only on increased knowledge, but also on prevailing tastes and the various purposes for which the maps are made and used. In both pictorial and literary depictions of London, for example, different centers are possible, and even where the center remains the same, different valorizations of that center produce alternative meanings, as we shall see. The center of London is not the same, for example, for monarchs as it is for peasants; likewise, the perceived center and the perception of the center of a particular country estate will certainly not be the same for the owner as for neighboring farmers.

Peripheries, margins, borders, and boundaries are also critical to this study. It will be important to note which details mapmakers and those who borrow mapping paradigms or spatial metaphors usher to the margins and which they exclude completely as they try to make sense of the world they see and to create a sense of place (whether at the local, national, or global level). Medieval mapmakers put monsters at the margins of their maps, or sometimes the warning "Beyond here there be dragons."[17] It also became customary to put the "monstrous races" that are described in traveler's tales at the edges of the known world and in the terra incognita.[18]

Biases and distortions in maps, both literal and figural, are also important to the current study. My purpose here is not so much to uncover or to amass an "accurate" view of the nation, the estate, or the city, as to uncover the biases that emerge with an eye toward understanding the basis of those biases. Mark Monmonier comments that, interestingly,

> Even folks who are routinely suspicious of written text equate maps with fact and fail to realize that no map is capable of including all information or telling all possible stories. In fact, the process of mapmaking requires cartographers to limit content in order to create a readable map and so allows them to manipulate their audience with the information they choose to include. (1)

We may have seen the humorous maps of a New Yorker's view of the United States in which Manhattan Island is nearly as large as peninsular Florida, the Midwest contains only a few sketchy squarish states, and San Francisco, California, and Hollywood are shown as separate states.[19] But all maps make choices about what to include and what to omit. In his consideration of various cartographic projections, David Turnbull suggests that

> The map, if it is to have authority in Western society, must have the appearance of "artless-ness"; that is, it must appear simply to exhibit the landscape, rather than to describe it with artifice or in accordance with the perceived interests of the mapmaker. For a map to be useful, it must of course offer information about the real world, but if this "real world information" is to be credible, it must be transmitted in a code that by Western standards appears neutral, objective and impersonal, unadorned by stylistic device and unmediated by the arbitrary interests of individuals or social groups. (1989, 8–9)

The map must "appear neutral," but Richard Helgerson reminds us that maps "could never be ideologically neutral" (1992, 147). In early modern maps that which is known is portrayed in much greater detail than that which is less familiar, or unknown, and the division between "here" and "there," between "home" and "away," becomes quite vivid. Perhaps the views of London that are the subject of chapter 5 will not appear as distorted as the example of the New Yorker's view of the United States, but Isabella Whitney's emphasis on the nether regions of the city is one example of a bias that is used to make a point, as we shall see. We are no doubt accustomed to seeing the prominently displayed kiosk map of the shopping mall, the university, or the subdivision that proclaims with great assurance: "you are here"[20]; this study will try to uncover where "here" is for the creator of the particular map or literary work, and what "here" means.

THEORETICAL AND CRITICAL BACKGROUND

Several works on the theory of "place" are valuable to my study. Michel de Certeau makes a useful distinction between space (*espace*) and place (*lieu*) in *The Practice of Everyday Life*. For him, a place "is the order (of whatever kind) in accord with which elements are distributed in relationships of coexistence. It thus excludes the possibility of two things being in the same location (*place*)." Space, on the other hand,

> exists when one takes into consideration vectors of direction, velocities, and time variables. Thus space is composed of intersections of mobile elements. It is in a sense actuated by the ensemble of movements deployed within it. Space occurs as the effect produced by the operations that orient it, situate it, temporalize it, and make it function in a polyvalent unity of conflictual programs or contractual proximities. (117)

Finally, for de Certeau, "space is a practiced place"; thus, for example, the geometry of urban streets "is transformed into a space by walkers" (117). This transformation is at the heart of the current study. Comparing space and place, de Certeau continues,

> in relation to place, space is like the word when it is spoken, that is, when it is caught in the ambiguity of an actualization, transformed into a term dependent upon many different conventions, situated as the act of a present (or of a time), and modified by the transformations caused by successive contexts. In contradistinction to the place, it has thus none of the univocity or stability of a "proper." (117)

Space thus becomes, to continue the reading analogy, "a place constituted by a system of signs."

Henri Lefebvre's *The Production of Space* (1991) links social and geographical issues in order to interrogate constructions of place in both creative literature and cartography. Lefebvre draws distinctions between physical space (perceived), mental space (conceived), and social space (lived), and then unites these distinct theoretical fields by postulating that social space, or space as directly lived, is a social product that gives meaning to place. Symbolic associations are produced and constantly change.[21] "Spaces" in de Certeau's analysis would be analogous to Lefebvre's "social spaces." What I find particularly compelling in Lefebvre and de Certeau's work is the idea that "space" is produced in the mind and by the interaction of social forces, rather than merely inhering in the fabric of the terrain, to engender what I would call "a sense of place." Maurice Merleau-Ponty sums up well the phenomenological apprehension of space to which both de Certeau and Lefebvre refer in various ways: "To experience a structure is not to receive it into oneself passively: it is to live it, to take it up, assume it and discover its immanent significance" (258); thus, objects or places perceived become invested by one's gaze, and become different for each gazer. My subtitle, "a sense of place," is meant to evoke this phenomenological notion of space.

Emphasizing the textuality of maps, cartographic historian J. B. Harley was among the first to highlight the importance of analyzing the rhetorical devices of maps, such as invocation, selection, omission, simplification, classification, and the creation of hierarchies (1987, 1–20). Consideration of these rhetorical devices is germane to my discussion of early modern mapping projects as well as to the other celebrations of place that I examine, such as the country house poem and the city comedy: thus, it is always important to consider what is omitted, what is invoked, and so on. Along similar lines, both J. B. Harley and David Woodward suggest that, like poetry, maps

> require painstaking interpretation in relation to their original purpose, their modes of production, and the context of their use. Maps created for one purpose may be used for others, and they will articulate subconscious as well as conscious values. Even after exhaustive scrutiny maps may retain many ambiguities, and it would be a mistake to think they constitute an easily readable language. Maps are never completely translatable. (3)

Finally, maps are used as a demonstration of power and authority. Whether it is the monarch commissioning a map of his/her kingdom or empire, a governmental advisor using such maps for strategy (or for pomp and display), a landowner wanting an accurate rendering of his financial holdings, or a city bolstering civic pride, maps are instruments of different sorts of power and authority. Equally important are the countercurrents to this type of authority; these countercurrents are evidenced by those who are *not* in positions of power but who nevertheless deploy maps and mapping paradigms to their particular purposes and thereby create, often in rather audacious ways, alternative models of power and authority.

Recent studies of mapping projects and geographic paradigms have broken important ground in examining the significance of mapping metaphors in the creative literature of the early modern period. Richard Helgerson's seminal *Forms of Nationhood: The Elizabethan Writing of England* (1992) examines England's evolution from feudalism into "nationhood" in terms of linguistic, legal, cartographic, and poetic discourses. His chapter on cartography, which investigates the political exigencies of map patronage and production, provides a context against which I read the textuality of early modern maps as one method of producing the "place of England." Frank Lestringant's *Mapping the Renaissance World* (1994) aptly presents the cosmographic changes that took place in the Renaissance, especially in the New World, and particularly in the work of André Thevet in Brazil. John Gillies's *Shakespeare and the Geography of Difference* (1994), exploring the connections between early modern geography and theater, provides an interdisciplinary model and tools with which to look at figurative renderings of maps, although his study, unlike my own, emphasizes world maps, constructions of the Other, and the theatricality of maps. Although his subject is French literature, Tom Conley's keen observations in *The Self-Made Map: Cartographic Writing in Early Modern France* (1996) are also apropos to British literature of the time period. Especially pertinent is Conley's definition of "cartographic writing." Conley deems various writings to be "'cartographic' insofar as tensions of space and of figuration inhere in fields of printed discourse" (3), and further contends that

> Many literary works of the years 1470–1640 appear to be seeking to contain and appropriate the world they are producing in discourse and space through conscious labors of verbal navigation. Writers borrow from a stock of geometric and cartographic commonplaces . . . to map out creations that are totalities much greater than their authors' own appreciation or conscious knowledge of them. (5)

Whereas J. B. Harley advocates examining the textuality of maps, Conley's study of "cartographic writing" examines the inverse of the equation, suggesting that:

> In incunabular and sixteenth-century literature, we behold works that betray the touch of the architect, the stage designer, the painter, and, no less, the cartographer. By virtue of spatial modes of composition, the writer tends to "map out" the discourse of the work before our eyes and to invite us to see the self constituting its being in patterns that move into space by means of diagrammatic articulations. (4)

Both approaches are useful to the current study.

Andrew McRae's *God Speed the Plough: The Representation of Agrarian England 1500–1660* (1996), with its thorough analysis of rural complaint, husbandry and improvement manuals, estate surveys, and poetics, and Garrett Sullivan's *The Drama of Landscape: Land, Property, and Social Relations on the Early Modern Stage* (1998), with its treatment of early modern landscape at the national, local, and metropolitan levels in the drama, have also been quite beneficial to my work. Raymond Williams's classic reading of English poetry *The Country and the City* (1973) and Immanuel Wallerstein's analysis of the economics of trade and empire in *The Modern World-System* (1974) have provided useful background as well. J. B. Harley's championing the rhetorical strategies used in maps and his examination of the textuality of maps has been foundational. Finally, the study of memory, especially of artificial memory systems, by Frances Yates in *The Art of Memory* (1966) and by Mary Carruthers in *The Book of Memory* (1990), has had a great impact on my thinking about maps and memory, as shown particularly in chapter 3.

Earlier studies of the intersection of maps and literature tended to focus on readings of physical maps that appeared in the literature or on metaphors of persons as maps: a map is brought on the stage in *King Lear,* an older man's face is referred to as looking like a lined map (Malvolio in *Twelfth Night*), a woman's body is described as being similar to a globe in which one could "find out countries" (Luce in *The Comedy of Errors*), and a lover's body is described as "America, my new found land!" (Donne, "Elegy 2"). My study, by contrast, looks at archival maps, contemporary surveys, and prose works about mapping and surveying techniques in early modern England and reads them in conjunction with more figurative evocations of maps in the literature: a disgruntled colonial secretary offers an array of figurative maps within an epic poem in order to convey his discontent with the state of foreign affairs, a Roman "surveys" an English princess and her chamber in the same manner that an Armada pilot would survey the coastline for penetrable places, a poet describes the details of the landscape and of the architecture of the estate as a surveyor might, and a female poet names and describes parts of London as a cartographer would. These are a few of the topics that I take up in the chapters that follow, where I uniquely address the pervasiveness of mapping metaphors and their use as more than merely another trope of description. My theoretical approach is both materialist and phenomenological, blending in elements of Marxist theory to investigate economic issues as well as feminist theory to interrogate issues of gender.

THE EARLY MODERN MAP IN ENGLAND

In 1579, Christopher Saxton's publication of his great collection of maps, *The Counties of England and Wales,* launched an interest in mapping and describing the terrain of early modern England that became something of a mania. County surveys, topographical maps, and atlases proliferated, and reproductions of maps decorated tapestries, paintings, and playing cards. According to Richard Helgerson, "for the first time [the English] took effective visual and conceptual possession of the physical kingdom in which they lived" (1992, 107). Descriptions of place—known as chorographies—also appeared in both poetry and prose.[22] William Camden's *Britannia* (1586), John Norden's *Speculum Britanniae* (1598), John Stow's *Survey of London* (1589), John Speed's *Theatre of the Empire of Great Britaine* (1611), and Michael Drayton's *Poly-Olbion* (1612) are a few of the better known works of cartography and chorography of the early modern period. Furthermore, the interest in the geography of England and its cities, villages, and landmarks resonates in creative works of poetry, prose, and drama of the time. This study focuses on the significance of representations of maps, of mapping activities, and of spatiality in the literature of early modern England and compares these linguistic representations to historical maps and to documents related to mapping activities.

Early modern ideologies about place and about cartography are quite different from those of the early twenty-first century. This point might seem to be obvious, but since maps up to early modern times had been owned and viewed primarily by those in positions of power, early modern societies as a whole were only beginning to be able to think cartographically. Conley's discussion of the innovative atlas speaks well to this point:

> The Ortelian atlas changes the way maps are to be viewed. Formerly, world maps and large sheets required a princely space in order to be

viewed from all directions. With Ortelius's portable folio atlas comes a miniaturization of the reading and viewing, which can be accomplished from any angle and from a distance that can be changed according to the way one walks around the book. Hence, the attention that the title page and the verso page draw to the mobility of the viewing subject confirms an intention to craft a "theater of operations" that can be imagined from all cardinal points, especially when the book is seen from the four sides of a table on which it rests. (216–7)

Many conventions of mapping that are now taken for granted, such as the *orientation* of maps toward the north, were just becoming established. As we have already seen, East is at the top of the T-O map and of the medieval *mappamundi;* this orient-ation carries over into some early modern maps, but other orientations are also attempted. "East" is also at the top of a Sebastian Münster map of England of 1540, but "West" is at the top of his 1578 map and on maps by Abraham Ortelius (of 1570, 1595, and 1601), and South is at the top of *Totius Britanniae tabula chorographica,* a map in a fourteenth-century edition of Geoffrey of Monmouth's *History of the Kings of Britain.*[23] Curiously, a mid-sixteenth-century house-plot in Durham has "North" at the top and "Sowthe" at the bottom, but it has "Est" [*sic*] on the left and "West" on the right; in other words, as P. D. A. Harvey suggests, the plan is not so much a mirror image as a view from the *ground up* rather than from above (1993 *Maps,* 9). Along the same lines, the now-familiar grid system used for locating and indexing places on maps by use of letter and number coordinates on the marginal axes was new enough in 1598 that John Norden's presentation copy of his *Speculum Britanniae* explains to Queen Elizabeth how to use the system that he had innovatively incorporated in his maps. Similarly, conventional iconic symbols for features such as cities, cathedrals, and hills were also not yet standardized.

In the remainder of this chapter, I examine some of the political ramifications of the innovative features of early mapping and survey projects: some maps are commissioned by Queen Elizabeth and seek to flatter her by elaborate descriptions of English landscapes alongside her coat of arms, some display the coats of arms of the local gentry rather than that of the monarch, some boast of exactness, some claim to correct the previous errors of others, and some purport to make travel easier for the uninitiated.[24]

One trend on which I focus throughout this study is the premium placed on correctness. Although not surprising in mapping projects that strive for increased scientific accuracy, the pursuit of correctness in these projects extends beyond the compass of the technical aspects of mapping enterprise: mapmakers as well as poets and dramatists seem to want to prescribe certain "correct" behaviors associated with particular places and they seem, at times at least, to have a moral design beneath what might initially seem to be a rather unbiased presentation and a morally neutral document. We have already noted that the maps themselves, while striving for neutrality, will always incorporate biases, but some ancillary cartographic materials warrant some scrutiny in this regard as well. For example, indexes to maps, usually assumed to be rather objective and unbiased lists, were often highly politicized in early modern England. John Norden's lengthy index of the places to be found on his maps of the *Speculum Britanniae* records, along with particular houses or places, incidents that took place at these locales and his own very telling comments about them. Consider the listing for "Bletchingleigh":

> At this place was a statelie howse belonging to the D.[uke] of Buckingham when he was apprehended of high treason for wordes against K.[ing] H.[enry]8. But some affirme his apprehension to be at the courte, being convented before the K.[ing] and that the wordes wer[e] uttered at his howse of Bletchingleigh to Georg Nevell Lo:[rd] Aburgavennye . . .
>
> (Norden, ca. 1595, unnumbered manuscript)

Queen Elizabeth would probably have appreciated this bit of local history regarding a late traitor to her father, especially since the house was by then a ruin.[25] Along these same lines, another Elizabethan bias is expressed in a list of houses that Norden sets out as "an alphabetical catalogue of the noble men and gent of accounte for the most parte and suche as in regarde of their welth seeme to usurpeth all title in Surrey w[i]t[h] their howses and present aboade w[i]t[h] in the same" (Norden). The latter part of this entry is a derogation of the much-derided "New Man" and expresses a bias against recently acquired mercantile wealth versus old aristocratic money and title. With or without Norden's editorial comments, such a list of wealthy persons (whether "noble men and gent of accounte" or usurpers) would be of paramount interest to the monarch.

As England emerged from feudalism, it became necessary for the monarch to be able to govern wider areas. Several writers of the early modern period address the expediency of maps in the functioning of government. Thomas Elyot in *The Boke Named the Governour* (1531), for example, writes of the efficacy of maps for allowing one to see countries that s/he could never visit, as he rhapsodizes as well on the pleasure to be had

> in one hour to behold those realms, cities, seas, rivers, and mountains, that unneth [scarcely] in an old man's life cannot be journeyed and pursued; what incredible delight is taken in beholding the diversities of people, beasts, fowls, fishes, trees, fruits, and herbs: to know the sundry manners and conditions of people, and the variety of their natures, and that in a warm study or parlour, without peril of the sea or danger of long and painful journeys: I cannot tell what more pleasure should happen to a gentle wit, than to behold in his own house everything that within all the world is contained. (35)

In *The Book of the Courtier*, Baldesar Castiglione endorses maps as a valuable aid to memory and to the presentation of one's plans: he points out that although a man may keep useful details of geography in his mind "yet can he not shewe them to others" without a map or painting (91). And this "shewing" is especially relevant to matters of state. For Castiglione then, maps act as physical props, but also as *aides-mémoire*. The kind of armchair travel envisioned by Elyot—the pleasure to be had from experiencing, even secondhand, the diversions of foreign lands—is similar to Norden's idea about estate surveying, that:

> a plot rightly drawn by true information, describeth so the lively image of a Mannor, and every branch and member of the same, as the Lord sitting in his chayre, may see what he hath, where, and how it lyeth, and in whose use and occupation every particular is, upon the suddaine view. (1618,[26] 15)

Availing oneself of a "suddaine view" from a "chayre" through the medium of a map comes in very handy for the exercise of power.[27]

The appearance of Elizabeth's portrait and/or her royal crest on the face of many maps suggests to those reading the maps her power over the country and the compass of her royal gaze—perhaps another version of the "suddaine view." In the Ditchley Portrait, Elizabeth's figure standing over a map of England looms so large that she nearly obliterates the map beneath, rendering it all but useless; on Christopher Saxton's frontispiece to his atlas, a portrait of Elizabeth still *covers* the maps and demonstrates her power over England, but here she serves as a dedicatory icon rather than an eradicating mantle (and she is joined by figures of cosmographers and geographers at her sides as well as beneath the larger image). Once this page is turned, it is possible to look at the maps without her ubiquitous presence, though perhaps her presence is always felt.[28] One might also recall portraits of Elizabeth with her hand on a globe, especially in the Armada Portrait, as demonstrations of her power over the map or the globe, even if illusory.[29] Lengthy and obsequious dedications to the Queen on some maps and atlases also serve to make her and the members of her court the definitive readers of these maps.

Elizabeth's chief minister, William Cecil, Lord Burghley, used maps for defense purposes such as fortifying the coasts in anticipation of a Spanish invasion, but he also used them for other governmental ends, such as locating recusant families and other potential rebels. In fact, the alphabetical indexes accompanying some of the maps of the Elizabethan period make it easy to locate the aristocracy for purposes of taxation and governance. Again, Norden lists members of the gentry in great number in the presentation copy of his *Speculum Britanniae* that is still preserved in Burghley's personal atlas in the British Library. Maps were also essential in locating justices of the peace, the local agents of Tudor government; in estimating tax revenues; in settling disputes; in estimating the number of men available for musters; and in planning travel routes for royal progresses.

Burghley was rather famous, in fact, for his collections of maps, and his distinctive annotations on the face of many surviving maps attest to his enthusiasm for the projects of mapping.[30] Indeed, Burghley decorated the galleries of his home with maps, including what was very likely a version of Christopher Saxton's large 1583 wall map of England that included not only all the cities, towns, and villages, but "also the armorial bearings and domains of every esquire, lord, knight, and noble who possesses lands and retainers to whatever extent," according to a report by Frederick, duke of Wittenberg (Rye 45). This type of display made it apparent, according to Peter Barber, that "even when contemplating the walls of his house, Burghley did not wish to escape, or wish his visitors to escape, from the concerns that occupied his mind when consulting his atlases and other works of reference in the solitude of his study" (Barber II, 77). Clearly, for Lord Burghley, printed maps and surveys facilitated both the exercise and the display of power.

Machiavelli, too, highlights the importance of topographical knowledge in *The Discourses* (III, 39), where he especially advocates hunting as a beneficial means for familiarizing oneself with the land. Moreover, he believes that the knowledge of topography gained through hunting

also enables one who has familiarized himself with one district, to grasp with ease the details of any new region. For all countries and all their parts have about them a certain uniformity, so that from the knowledge of one it is easy to pass to the knowledge of another. . . . A person who has had practice, for instance, will see at a glance how far this plain extends, to what height that mountain rises, where this valley goes, and everything else of this kind, for of it all he has already acquired a sound knowledge. (511)[31]

By extension, one can see how the counties of England, though different, can become known through comparing similar features from one county to another, and from one map to another. If we apply Conley's concept of "cartographic writing" here, we might add that understanding the uniform iconic symbols used on maps adds to the skill that Machiavelli advocates, that "enables one who has familiarized himself with one district, to grasp with ease the details of any new region" (511).

The royal progress is another means of royal survey, one in which the monarch visits the geopolitical subdivisions of his or her kingdom in person. Machiavelli advocates this kind of firsthand knowledge in his *Art of War* (1521). Machiavelli's advice is given in the context of war, but his advice is nevertheless apropos for the monarch on progress. He points out that in order to avoid exposure to the dangers of marching through an enemy's country,

The first thing [a general] ought to do is to get an exact map of the whole country through which he is to march so that he may have a perfect knowledge of all the towns and their distance from each other, and of all the roads, mountains, rivers, woods, swamps, and their particular location and nature. (143)

Machiavelli suggests that one should get these maps from several sources, question those providing the information, compare their accounts, and then send one's own cavalry parties to verify the accuracy of the maps.

At the more local level, separate county maps facilitated visual apprehension for the local leaders, sheriffs, and justices of the peace. Coats of arms of the local gentry, used as marginal ornamentation, emblematized their dominion in the same way the queen's crest on national maps made her presence known. Local maps could also be used by local powers for defense—locating beacons or preparing musters.[32] Even at the most local level—the estate map—the urge to display power and dominion is apparent in the family crests and family trees that serve as marginal, but significant, decoration.[33]

A number of marginal and subsidiary mapping conventions contributed to the display of power and the impulse to "contain" via mapping: not only map indexes, but also essays about mapping and dedications to atlases worked together to prescribe and proscribe certain behaviors. The persona of the surveyor in Norden's *Surveiors Dialogue* fancies himself to be a person entitled to dispense or codify correct social behavior on matters as diverse as how to address a Lord and how to enrich the soil. Elsewhere, in his indexes to the various maps of his *Speculum Britanniae,* Norden does not stop with merely locating a particular place on the map by use of coordinates on marginal axes; oftentimes his index entries contain rather didactic information about events that transpired at those places (from which, presumably, readers are to take their cues as to how to behave). The listing for "Bletchingleigh" above is one example in which the information Norden supplies serves to fix that place and its lesson in the reader's memory.

In addition to helping one familiarize oneself with the landscape and locate particular features and people, maps also facilitate remembering specific details about the country. In use since classical times by rhetoricians, artificial memory systems rely on some of the same spatializing and compartmentalizing activity that occurs in mapping the subdivisions of a geopolitical area. Especially in times before the widespread use of print, or of paper for taking notes, artificial memory systems were popular devices for systematizing, visualizing, and memorizing large bodies of information. Mary Carruthers suggests that memory systems could take the form (in the mind) of nesting places for small animals, dovecotes, bee hives, pigeon holes, Noah's Ark, or a money purse and coins (Carruthers 35–39). Using a memory system, materials to be remembered are arranged mentally into the places (or *loci*) of whatever organizing system is selected, be it beehive, dovecote, or Ark. Some of the most popular systems in the early modern period were called "memory theaters."[34] In the memory theater the viewer occupies a central position and visualizes the material to be memorized in an architectural form in front of and around him or her. The architectural form of the typical memory theater is imaginary, but it may be based on a real building, such as a house or a palace. The best architectural forms have an abundance of niches, columns, rooms, arches, doors, and windows so that each bit of information to be memorized can be assigned a place somewhere within this architectural form in the memory. This schema allows one to retrieve the memorized information by visualizing the architectural structure and then recalling the material situated within each location.

Frances Yates traces the history of artificial memory and of memory theaters in *The Art of Memory*, beginning with a story that Cicero narrates of Simonides identifying the crushed bodies of victims of a collapsed roof by means of remembering where each person was sitting at the banquet during which the roof collapsed. According to Cicero, Simonides

> inferred that persons desiring to train this faculty (of memory) must select places and form mental images of the things they wish to remember and store those images in the places, so that the order of the places will preserve the order of the things, and the images of the things will denote the things themselves, and we shall employ the places and images respectively as a wax writing-tablet and the letter written on it. (*De oratore*, II, lxxxvi, 351–4, qtd. in Yates 2)

Cicero emphasizes that Simonides' invention of the art of memory rested not only on his discovery of the importance of order for memory, but also on the discovery that the sense of sight is the strongest of all the senses and "that consequently perceptions received by the ears or by reflexion can be most easily retained if they are also conveyed to our minds by the mediation of the eyes" (*De oratore*, II, lxxxvii, 357, qtd. in Yates 4).

Similarly, an unknown rhetoric teacher between 82 and 80 B.C. compiled some of the fundamental elements of the art of memory, in a work now known only as *Ad Herennium*.[35] The first step in this artificial memory system is to form a set of *loci* upon which to imprint the images to be remembered. These *loci* can be used again and again for remembering different sets of information or different orations. Yates stresses that "the loci are like the wax tablets which remain when what is written on them has been effaced and are ready to be written on again" (7). In rhetoric, the *loci* can be used for remembering different speeches. In cartography, different maps are imprinted over the same terrain, but may contain different bits of information depending on the purpose for which the map was drawn. We might think of the various purposes for which maps are drawn or the kinds of geography that are represented on maps: topographical, economic, human, political, physical, historical, and so on.

In the "theater of memory" designed by Giulio Camillo in the sixteenth century, a theater is spread out in front of the viewer, who occupies center stage.[36] Each aisle, each row, each tier, and each compartment is then filled up with the information to be recalled. Dividing up the map of England into its constituent parts (counties, hundreds, townships, parishes, rapes) allows one to view the map at once as a memory theater (an *aide-mémoire*) but also as a replacement to memory during a time when written and printed materials were becoming more widely available. Likewise, one can visualize a map and retrieve vital information and one can "re-member" the whole by reassembling the parts. Most of us probably learned the subdivisions of our own country in a similar visual way. Not coincidentally, Abraham Ortelius's first atlas, in 1570, was entitled the *Theatrum Orbis Terrarium,* the "Theater of the World"; and many succeeding atlases also included the word "theater": John Speed's *Theatre of the World* and his *Theatre of The Empire of Great Britaine* are two contemporary examples. Fittingly, too, theaters often performed some of the work of maps, by gathering and presenting peoples and scenes from various parts of the world—one thinks particularly of the "Globe Theater" in this regard.[37] Maps were used to enhance the knowledge and domination of those already in power; but mapping paradigms and other figural portrayals of place serve poets, dramatists, and prose writers who wish to convey alternative statements of ascendancy.

Overview of Chapters 2–5

The chapters that follow delineate a progression of cartographic writing in early modern England that begins at the national level and continues to the local level. My next two chapters deal with the subject of national representations of England from the viewpoint of outsiders as well as insiders. Chapter 2, "Marriage Pageants and Ceremonial Maps," centers on native mapping activities and the uses made of maps at court. The focus of my discussion here is Edmund Spenser's *The Faerie Queene* (1596). I argue that in the episode of the Marriage of the Thames and Medway Rivers (Book 4, canto 11), Spenser is presenting a ceremonial pageant of maps (ranging from ancient, to medieval, to early modern), in order to curry favor with the queen for his pet colonialist projects in Ireland and the New World. I argue that contemporary mapping and surveying projects such as Christopher Saxton's and John Norden's mapping of the counties of Britain, the extensive mapping and surveying of Ireland, the sometimes quite imaginative mapping of the New World with an eye to colonization, and the more practical estate maps are exemplified within this episode, and that Spenser uses figurative maps not only to flatter the queen but also to censure her (if only indirectly) for a foreign policy with which he disagrees. Writing in the mid-1590s, with succession concerns deepening, Spenser also seems to suggest that the queen's body, like some of the outmoded mapping paradigms he presents, is outdated. I present a reading of parts of *The Faerie Queene* against an expanding world view, the great English mapping projects of the day, Lord Burghley's avid interest in maps for strategic purposes as Elizabeth's secretary of state, and early modern monarchs' use of maps as means of propaganda and display.

Chapter 3, "Maps, Figures, and Figurative Maps: Feminine Geography," focuses on figurations of the topography of Britain and the New World as a feminine territory, and analyzes surveying and mapping projects conducted by outsiders considering invasion and conquest of the these feminized territories. My discussion focuses on Shakespeare's *Cymbeline* (1611), a play in which England becomes, metaphorically, a feminine landscape to be surveyed and mapped. Here, the Roman Jachimo secretly surveys the bed chamber of the sleeping British princess Imogen in a manner similar to that of a cartographer. Using his keen memory, he is able to present the evidence he has gathered to Imogen's husband, Posthumus, as proof that he has explored Imogen's body as well as her chamber—in short, that he has enjoyed a sexual encounter with her. I discuss Jachimo's charting of Imogen's terrain in light of early modern mapping projects, but also in terms of artificial memory, and I argue that the initial evidence that Jachimo presents to Posthumus is like a portolan, or coastal, map in which the penetrability of Imogen's body and of her chamber reflects England's uneasiness about its own vulnerability to invasion. But in this particular case maps are inadequate to the purpose at hand, as we shall see.

I turn in chapter 4, "Landscape, Labor, and Legitimacy," to more local representations as I discuss estate surveying. P. D. A. Harvey suggests that Saxton's maps of the counties of England, having introduced the English gentry to the use of the scale map, made it "easy, or at least possible, for the enterprising surveyor to sell the landowner the idea of having a scale-map of his own estate" (1993 "Estate Surveyors" 45). Here I investigate works (such as John Norden's *Surveiors Dialogue* [1607, 1610, and 1618] and Radolph Agas's *Preparative for Plotting of Lands and Tenements for Surveigh* [1596]) that detail the techniques and the advantages of these surveys in conjunction with the verbal "survey" reflected in a group of poems celebrating the aristocratic lifestyle of the country residence, known as the genre of the "country house poem." In Ben Jonson's "To Penshurst" (1616), for example, the poet describes the aristocratic house and the grounds as a means of praising its owner, while inhabiting a position similar to the estate surveyor: someone who is not *of* the aristocracy but who is allowed to "measure" the land and appraise the lives of aristocrats while in their employ. The country house poem is one genre that has not been addressed adequately for its geographic content, but the twin genres of the estate survey and of the country house poem, in vogue at the turn of the seventeenth century as celebrations of place and ownership, enter into a shared system of representational practices. I investigate the impetus for these poetic and cartographic celebrations of place, nobility, and ownership that gain popularity at a time when the foundations of the aristocracy were beginning to be questioned. I also examine how the country house poem evokes and reflects pride of ownership, how it glorifies (but sometimes slyly castigates) doctrines of virtue and good stewardship, how it accounts for gender issues, and how it reflects the discovery and colonization of the New World through its frequent paradisiacal allusions.

Turning to metropolitan quarters, I focus chapter 5, "Cityscapes and City Scrapes," on city maps and urban representations. I highlight literary works in which the descriptions of the streets and districts of London and its busy denizens resemble early modern maps of London such as the Woodcut Map, the Copperplate Map, and the first printed map of London by Georg Braun and Fran Hogenberg. Toward that end, I discuss Elizabeth's coronation progress (1559), Isabella Whitney's "Wyll and Testament" (1573), John Stow's *Survey* (1598 and 1603), and the genre of the city comedy, all of which pay particular attention to the demarcation of distinctive districts. The coronation progress endeavors to elevate the city's religious and commercial districts while Whitney's "Will" and city comedy tend to caricature and lampoon them. It is not simply the topical references to streets, buildings, and landmarks that bring these works into the realm of cartographic and chorographic inquiry, but the ideologies of place that they contain; places, like maps, are never neutral, but are always filled with meanings—culturally shared meanings as well as meanings based on personal experiences.

This study concerns itself with both the "lines in the map" and the "maps in the lines" of poetry, drama, and prose. Cartographers, surveyors, and poets use cartographic representations in myriad ways. The locations depicted are value-laden: sometimes pure, untainted, and virginal; sometimes enticing; sometimes ennobled, revered, and even saintly; sometimes wanton and incontinent; and sometimes predatory and devouring. Reflecting this multiplicity, we shall see that writers and mapmakers have various relationships with these places: some want to portray them "accurately" (according to scientific methodology), others want to contain, control, or own them, still others want to correct unruly behavior, or to use maps as an avenue of correction for society's ills and deficiencies. J. B. Harley reminds us that

> Maps are never value-free images; except in the narrowest Euclidean sense they are not in themselves either true or false. Both in the selectivity of their content and in their signs and styles of representation maps are a way of conceiving, articulating, and structuring the human world which is biased towards, promoted by, and exerts influence upon particular sets of social relations. By accepting such premises it becomes easier to see how appropriate they are to manipulation by the powerful in society. ("Maps" 278)

These "sets of social relations" are the fertile ground for the current examination of maps, memory, and representation in early modern England.

CHAPTER TWO

MARRIAGE PAGEANTS
AND CEREMONIAL MAPS

THE MARRIAGE OF THE THAMES
AND THE MEDWAY
(SPENSER'S *Faerie Queene* BOOK 4, CANTO 11)[1]

In the late sixteenth century, as we have seen, both Thomas Elyot and
Niccolo Machiavelli ardently advocate the use of maps by governmental ministers. Machiavelli highlights the importance of topographic and
cartographic knowledge as an aid to military commanders and Elyot's
Boke Named The Governour—providing guidelines for the education of
future leaders—considers the importance of maps for the business of
government as well as for personal pleasure. Elyot rhapsodizes, as I discussed in chapter 1, on the pleasure to be had in beholding "everything that within all the world is contained" without leaving one's
home (35). Similarly, Baldesar Castiglione sees maps as a valuable aid
to memory and to the presentation of one's plans; in his *Book of the
Courtier* he points out that although a man may keep useful details of
geography in his mind "yet can he not shewe them to others" without
a map or painting (1967, 91). This "shewing" is especially relevant to
matters of the court. Perhaps because they are instruments of both
statecraft and pleasure, maps were popular courtly devices. And even
in a time of increasing availability and greater familiarity with more
modern maps that had practical applications in defense, land law, and
discovery, it is interesting to note that modern as well as outdated maps
held sway for ceremonial purposes. The presentation of maps was a

customary form of political compliment and a means of currying favor at court. Peter Barber elucidates how subjects sought to influence royal policy through the presentation of maps,[2] and he points out that "maps and plats were not objects like books and tapestries that might be found anywhere. By 1550 they were particularly associated with places intended for pageantry, propaganda, and government" (42), such as Whitehall, Greenwich, Hampton Court, and St. James Palace. Such displays included maps of current domains and recent conquests, as well as older world maps, or *mappaemundi*—a use which "stemmed from the *mappamundi's* role in medieval society as an expression of history, religion, and legend as well as of geography" (Barber I, 26).[3] In these, as well as broader contexts, maps were becoming well enough known that we can talk about a nascent "map consciousness," as do Peter Barber, Victor Morgan, and P. D. A. Harvey (Barber II, 58).[4]

Given the importance of maps for ceremonial and political purposes, it is not surprising that Edmund Spenser would present a series of maps, translated into ecphrastic poetry, in the specific moment of *The Faerie Queene* with which this chapter deals, the River Marriage of Book 4, canto 11. In this chapter, I argue that, as in courtly pageantry, Spenser presents figurations of antique maps along with newer ones—such as those from different time periods that were displayed together in Whitehall and Hampton Court Palace, and those that might be displayed together in a map room, study, or *studiola*. More specifically, I argue that Spenser begins his presentation with a paradigmatic medieval *mappamundi* as a ceremonial (but ambivalent) gesture to Elizabeth and as a prelude to the pageant of rivers to follow—a pageant that blends both antique and early modern geographical notions and also incorporates a broad spectrum of poetic figurations of the maps and cartographic projects that were in vogue in Spenser's day. I will argue that Spenser employs this pageant of ecphrastic maps to convey his dissatisfaction with Elizabeth's foreign policy (particularly in current and potential colonies) and to impugn her for her failure to produce an heir or to name a successor, two powerful messages that he might not be able to convey safely in a less subtle manner. Both of these criticisms are brought to culmination in his suggestion, cloaked in cartographic description, that Elizabeth's legacy depends on her letting go of antiquated notions of geography and choosing more modern maps that include the New World. Spenser's geography both abroad and at home is permeated with a discourse of rape that associates notions of New World conquest with unease about the invadability of England as a country and of the queen, its sometimes metonymic equivalent.[5]

At the outset of the river marriage episode, Spenser accords a few stanzas to the fate of Florimell, held captive by Proteus. Other scholars have devoted little attention to these few stanzas and, in particular, none has found a way to integrate them with the marriage pageant. I would like to suggest that by viewing the canto as a whole in the light of cartography, the first few stanzas not only come into focus themselves, but shed new light on the rest of the canto. In such a light, the Florimell stanzas can be seen to relate to Spenser's larger purpose rather than being a mere update on the fate of this perennially distressed maiden. I want to argue that with his description of Proteus's cell that imprisons Florimell, Spenser employs a curious geographical representation that has elements of both the ancient "T-O" map and of medieval geographic myths in order to facilitate a move into his own early modern cartographic vision later in the canto. This vision includes insuring England's place on the map through greater attention to colonial administration and expansion.

As we saw in chapter 1, in the conventional T-O map, the land mass is surrounded by the ocean (the "O"; though this "O" can take shapes other than strictly round) and is divided into the three known continents by other major bodies of water that form a "T." According to John Noble Wilford, "The style [of the T-O map] was symbolic, ornamental, and often beautiful; the geographic content, impoverished and usually misleading; the purposes, a representation of the mind more than of the Earth" (Wilford 45). This ancient schema evolved into the more detailed, but still highly symbolic, medieval maps that retain the chief elements of the T-O; that is, they continue to feature an encircling ocean and prominent bodies of water in the center that preserve a semblance of the T configuration (but with added detail), the positions and relative proportions of the three continents remain the same, east continues to be at the top, and Jerusalem remains at the center.

The river-marriage canto begins by reminding us of the treachery of Proteus's plan to "compell [Florimell] by crueltie and awe"[6] in stanza 3's dungeon "walled with waves":

> Deepe in the bottome of an huge great rocke
> The dongeon was, in which her bound he left,
> That neither yron barres, nor brasen locke
> Did neede to gard from force, or secret theft
> Of all her louers, which would her haue reft.
> For wall'd it was with waues, which rag'd and ror'd
> As they the cliffe in peeces would haue cleft;

> Besides ten thousand monsters foule abhor'd
> Did waite about it, gaping griesly all begor'd.
>
> (4.11.3)

Here, Spenser's description of Proteus's "dongeon" mirrors the encircling waters of both the ancient T-O map and the successor medieval *mappamundi*,[7] adding the medieval element of decorating the
map with monsters "foule abhor'd" who lurk in the margin "gaping
griesly all begor'd."[8] Spenser's descriptions follow the pictorial convention of decorating the margins of maps with monsters, as seen on
the Psalter and the Ebstorf maps, where dragons embellish the borders and all the "monstrous races" are located at the edge of Africa
(on the right side of these maps; see Figures 1.2 and 1.3). However,
whereas both the T-O map and the *mappamundi* customarily locate
Jerusalem at the center of the world, Spenser's little island map has
at its center a likeness of the pagan underworld of stanza 4 where the
river Styx lies, and light never shines:

> And in the midst therof did horror dwell,
> And darknesse dredd, that neuer viewed day,
> Like to the balefull house of lowest hell,
> In which old *Styx* her aged bones alway,
> Old *Styx* the Grandame of the Gods, doth lay.
> There did this lucklesse mayd seuen months abide,
> Ne euer euening saw, ne mornings ray,
> Ne euer from the day the night descride,
> But thought it all one night, that did no houres diuide.
>
> (4.11.4)

Spenser calls attention to the Styx's centrality to this image of
Florimell's prison by placing the river at the center of the stanza (lines
4–5) and at the center of the dungeon that holds Florimell; with the
same maneuver, the "hell" of Florimell's vagina (another locale of
"darknesse dred, that neuer viewed day") also becomes central to the
"map" and to the stanza.[9] The walls of waves as well as the "ten
thousand monsters" of the previous stanza, which guard against any
"force, or secret theft" with which Florimell's lovers "would her have
reft," make nearly palpable a fear of invasion and rape inherent in
Spenser's "map."[10] Like Jerusalem in the T-O map and the *mappaemundi* (along with the metaphoric association of Christ's navel in the
Psalter and Ebstorf maps), Florimell (and the rapeable "hell" of her
vagina) compels the action of the stanzas in which she is held captive.
Just as Leonardo replaced Christ's navel of earlier maps with the male

genitals in his drawing, in this secular and descriptive map with a woman's body at the center, the female genitals occupy the omphalos position and thus act, in the same way as Christ's navel (and his nativity), as a centering device.

The *isolario* map is another cartographic genre that Spenser may be representing here. Founded in the fifteenth century by Christopher Buondelmonti, this genre of island maps addressed, according to Frank Lestringant, "a clientele of the sedentary" and is "little concerned with precision or with providing the professional user with reliable information" (1994, 109). Buondelmonti's *Liber Insularum arcipelagi* (1420), depicted the islands of the Aegean archipelago and retraced legends of classical mythology. Other works in this genre were Bartolommeo dalli Sonetti's *Isolario* (1484), Benedetto Bordone's *Libro . . . di tutte l'isole del mondo* (1528 and later editions), Tommaso Porcacchi da Castiglione, *L'isole piu famose del mondo* (1572), André Thevet's *Grand Insulare et Pilotage* (1586), and finally Vicenzo Coronelli's *Isolario dell'Atlante Veneto* (1696).[11] If Florimell is not exactly imprisoned in a map, her prison within Proteus's cell is mapped in the poem as an *isolario* map of the style that was popular in the early modern period.

But Spenser's representation here recuperates outdated notions besides those of the medieval maps. It also revisits some tired old masques put on for the queen's entertainment many years earlier,[12] the Arthurian legend of the Lady of the Lake, and Queen Elizabeth's previous imprisonment in the Tower of London when she was held, like Florimell, in a "rock" surrounded by waves—since the Tower was, until the 1840s, surrounded by a moat. If presenting a ceremonial map is meant to compliment his reader, the queen, the subtext of imprisonment seems to undercut that compliment. But contradictions of this nature are not all that unusual in courtiers' dealings with Elizabeth. Susan Frye points out that

> By the 1570s and 1580s, as the queen and those who competed with her for representation produced a complex iconographic system, Elizabeth I's perfections and limitations were increasingly defined through the motif of imprisonment and delivery. The result was a dynamic expressing her courtiers' *fantasies of defining and controlling her.* (1993, 77, emphasis added)

This type of control seems to fit Spenser's agenda perfectly. What better way to get his monarch to attend to his own opinions than to hold her prisoner for a while—both in a poetic reminiscence of her

own imprisonment, and in the figurative substitution of Florimell?[13]
As Frye demonstrates, representations of Elizabeth's imprisonment
exalted her courage while participating "in one of the most prevalent paradigms for the containment and subordination of women"
(77). Although Spenser's narrator claims to feel some "pittie" for
Florimell's predicament, he begins the canto by reminding us that it
is, after all, through *his* agency that Florimell has been languishing
so long:

> Bvt al for pittie that I haue thus long
> Left a fayre Ladie languishing in payne:
> Now well away, that I haue doen such wrong,
> To let faire *Florimell* in bands remayne,
> In bands of loue, and in sad thraldomes chayne;
> . . . .
> That euen to thinke thereof, it inly pitties mee.
>
> (4.11.1)

The narrator's contrition seems disingenuous here, since he is the
one who has left Florimell in this "thrall" for nearly seven thousand
lines, and since he will soon leave her "languishing in payne" for
nearly five hundred more lines before even mentioning her again.[14]

The language Spenser uses in describing Florimell's imprisonment
resembles that used by John Foxe in *Actes and Monuments* (originally
published in 1563)[15] to describe Elizabeth's imprisonment; after
being "clapped up in the Tower" for some time, she was then

> tossed from thence, from house to house, from prison to prison, from
> post to piller, at length also [she was] prisoner in her own house, and
> guarded with a sort of cut throtes, which ever gaped for the spoile,
> whereby they might bee fingering of somewhat. (1895)

Both the sense and the language here are similar to Spenser's monsters
"gaping griesly all begor'd"—and certainly the desire of these "cut
throtes" to "bee fingering of somewhat" reflects the topos of rape.
Similar, too, is the language of George Gascoigne's masque presented
at Kenilworth in which Iris asks a queen named "Zabeta": "Were you
not captive caught? were you not kept in walles? / Were you not forst
to leade a life like other wretched thralles?" (2.5.21–22); Spenser may
be directly echoing Gascoigne with his phraseology of the "dongeon
wall'd with waves" and "thraldome's chayne." For Foxe, the eventual
freeing of the Princess Elizabeth is dependent on "the admirable working of Gods present hande in defending and delivering her" (1895);[16]

Figure 2.1 Ditchley Portrait of Elizabeth I, Marcus Gheerhaerts the Younger, 1592
(By courtesy of the National Portrait Gallery, London)
 Elizabeth stands on a map of England, demonstrating her metonymic association
with the country and the island.

her situation is mirrored in Florimell's imprisonment, "from which vn-
lesse some heauenly powre her free / By miracle, not yet appearing
playne, / She lenger yet is like captiu'd to bee" (4.11.1.6–8). That
Spenser's diction in this episode is so similar to these two other accounts regarding Elizabeth, with which he would have been familiar,
reinforces the connection he seems to be making between Florimell's
current circumstance and Elizabeth's former imprisonment, and more
generally between the persons of Florimell and Elizabeth. By such a
connection, Florimell becomes yet another of Spenser's "mirrours
more then one" for the queen (3.Proem.5).

In the pictorial realm, Florimell's circumstance may remind us of
Elizabeth's stance in the familiar Ditchley Portrait, where queen and
country seem almost one. While the queen's image dominates this
iconography and obscures much of the map on which she stands, we
cannot help noticing the tiny ships that sail beneath her skirts into the
Bristol Channel and the Thames Estuary (Figure 2.1). The rapeable
vaginal "hell" such as that at the center of Florimell's dungeon/map
is also suggested in this portrait. These ships represent the dual possibility of military invasion or commercial enterprise, both of which
can be read as unacceptable transactions with a chaste queen (the one
representing rape, the other prostitution) in this iconography and in
a culture that is so disposed to represent its own vulnerability to invasion as a threat to the queen's "virgin knot."[17]

Florimell's imprisonment in a "rock" surrounded by waves may
also allude to England's island status. In the nomenclature of the
day, seas and oceans had not yet been completely distinguished
from each other. Although it sounds odd to modern ears, Columbus had been, only a century earlier the "Admiral of the Ocean
Sea." In Christopher Saxton's map of England of 1579 the seas
surrounding Britain are still called "oceans": *Oceanus Britannicus*
and *Oceanus Germanicus* for the British Sea (later the English
Channel) and the North Sea, respectively; only the Irish Sea is
called a "sea": *Mare Hibernicum*. By the time of John Speed's atlas
in 1610, these bodies of water have become "The British Sea,"
"The Germain Sea," and the "Irish Sea"; the Atlantic Ocean is
now called "The West Ocean." With England surrounded primarily by "oceans" rather than seas, Saxton's depiction seems like a little world or a T-O map all to itself. As well, references to the seas
surrounding England as both a moat and a wall are commonplace,
as in Shakespeare's John of Gaunt's later reference to England as a
"fortress built by Nature for herself," emphasizing the protective
properties of the encircling seas:

This precious stone set in the silver sea,
Which serves it in the office of a wall,
Or as [a] moat defensive to a house,
Against the envy of less happier lands.

(*Richard II*, 2.1.46–49)

Later in the same passage, Gaunt describes England as "bound in with the triumphant sea, / Whose rocky shore beats back the envious siege / Of wat'ry Neptune" (61–62). Here it seems as if the sea assaults and invades rather than protects; and it is the *shore* that offers protection from the envious siege of invaders, such as Spain, that might arrive by sea. Presumably, too, Florimell's lovers and most of Elizabeth's suitors would arrive by the same medium.

As Spenser once again abandons Florimell after only seven stanzas, he simultaneously turns away from the representations of ancient and medieval geography that seem to imprison her (and by extension Elizabeth). He turns next to sea gods and to founding myths of nations and then to more specific and increasingly more contemporary vistas. The six cantos that Spenser devotes to the gods of the seas and the founders of nations add some detail to sketchy creation myths from Ovid and from Genesis where the land is simply (or miraculously) separated from the water. Spenser's "watery gods" of stanzas 11 through 14 control the seas that must be crossed and mastered by navigation in order to enable the founding of nations represented in stanzas 15 and 16, especially of Albion, which of course has special significance among all the nations represented here. Admitting the vastness of the project Spenser has undertaken, however, by stanza 17 he realizes that he must relinquish some details to those "that are better skild, / And know the moniments of passed times" (antiquarians, in other words); henceforward he will limit himself:

Onely what needeth, shall here be fulfild,
T'expresse some part of that great equipage,
Which from great Neptune do deriue their parentage.

(4.11.17)

Thus, after only two more stanzas narrating the progress of Ocean, his Dame Tethys, and their son Nereus, Spenser is ready for the pageant of rivers.

River marriage is not original with Spenser, but in my reading he uses it in a more (subversively) political way than his predecessors had. John Leland's *Cygnea Cantio* written in Latin in 1545 describes a journey of swans "reviewing the beauties and antiquities of various sites down along the Thames," including Reading, Windsor, Eton, Richmond, Kew, London, Greenwich, and Deptford. Then, after praising England's ships and its sea power, it eulogizes Henry VIII. Finally, the swan bids its mates farewell, "in preparation for its journey to heaven" (Roche 171). This poem, classified by Thomas Roche as a river marriage, might be considered a forerunner of Spenser's river marriage even though, as Jack Oruch points out, there's no actual river marriage here (Oruch 613). William Camden's fragmentary *De Connubio Tamae et Isis* scattered throughout his *Britannia* was published only a little at a time in the six early editions between 1586 and 1607.[18] William Vallan's *Tale of Two Swannes,* published in 1590, has also been suggested as an antecedent to Spenser's river marriage. None of these three, according to Roche, shows the depths of symbolism present in Spenser's poem, however. Finding it difficult to generalize based on these few poems, Roche concludes only that "the river marriage was employed as a device by learned Court poets, [and] that it was used as a frame to unite political praise with descriptions both historical and geographical" (Roche 173–4). In Spenser's river marriage, however, as we shall see, that expectation of "political praise" is often subverted.

In Spenser's creation, rivers and towns become transformed by what has happened there: these "places" on the map become the "spaces" of the poem. Thus charged with multiple and ambiguous meanings, even real locations are perfect for Spenser's mythical ceremonial purposes. In particular, Spenser's choice of the Thames as one of the rivers for this nuptial pageant operates on several levels of memory. First, it reiterates the earlier allusion to Elizabeth's imprisonment in the Tower, a prominent feature of the Thames at London. Next, it foregrounds the vulnerability of this queen and of a number of her castles and palaces along the river—specifically, Greenwich, Windsor, the Tower of London, Hampton Court, Syon, Whitehall, and Oatlands. Finally, it calls attention to the passageway to the city of London, and to England as a (w)hole, through the Thames estuary. This estuary, which functions as a symbolic geographical vagina to the body politic of England, and which was crucial in guarding the waterway to London, was constantly fortified against enemy attack.[19] Spenser's choice of the Thames is complicated and ambiguated by the fact that, despite the vulnerability this river represents, it also summons up recollections of the defeat of the Spanish Armada at the hands of the military under the command of Queen Elizabeth and of her rousing speech to the land forces at Tilbury, near the mouth of the Thames, in anticipation of the Armada attack. But it is important to remember that England's fear of invasion did not disappear even after the defeat of the Armada in 1588, nor did Spain's interest in a successful invasion.[20]

Likewise, the choice of the Medway has both military and poetic connections: it was the base of British naval operations, and it was frequently associated with the late Sir Philip Sidney, the quintessential Renaissance gentleman, poet, and sometimes court favorite, whose family home of Penshurst Place is along the banks of Medway in Kent.[21] Spenser, in fact, has paid tribute to Sidney in several of his other poems: in the *Ruines of Time*, Spenser refers to Sidney as the "worlds late wonder, and the heavens new joy" (303);[22] in *Colin Clouts Come Home Againe*, Sidney is figured as "Astrofell" who had no "Paragone" in his lifetime (449–451). But Spenser is critical of Sidney in his elegy "Astrophel," where his narrator cannot understand the desire that Astrophel, a shepherd, has to "hunt" (that is, to fight in battle), particularly in the foreign soils where Sidney lost his life (79–120).[23] Moreover, the association with Sidney may be less than agreeable to the queen because of Sidney's banishment from her court for protesting her expected marriage to the French Duc d'Alençon.[24]

But presenting Elizabeth with a marriage pageant at all, even one saturated in the geographical allegory of the Thames and Medway rivers, seems a bit risky (or at least outdated and quixotic) in 1596, when Books 3–6 of *The Faerie Queene* appeared. As early as 1566, Elizabeth expressed her irritation with her Parliament and her subjects who desired her marriage but who seemed unwilling to accept any of her potential suitors: "they (I thynke) that mouythe the same wylbe as redy to myslyke hym with whom I shall marrie, as theye are nowe to move yt, and then yt will apere they nothynge mente it."[25] Broaching the subject of her marriage (and of succession) had famously dire consequences not only for Sidney, but also for John Stubbs and Peter Wentworth.[26] Not surprisingly, then, it seems that Elizabeth censored a wedding masque in Robert Dudley's Kenilworth entertainments of 1575 after she had already had her patience tried with a "bride dael" featuring an aging, "ill smelling" bride that, as Frye points out, mocks Elizabeth as an unmarried middle-aged woman (1993, 62). This masque featured Iris and Diana debating the respective merits of virginity and marriage in front of a queen playing a nymph with the barely disguised name "Zabeta." Sidney makes similar gestures to Elizabeth in *The Lady of May* and *The Four Foster Children of Desire*.

The idea of courtiers performing their obeisance to the queen in tournaments and, especially, in masques is imitated, I think, in Proteus's presentation of himself to Florimell earlier in *The Faerie Queene*—as he "entertained her the best he might . . . To winne her liking vnto his delight" (3.8.38)—as a series of wooers: a God, a "mortall," a "Faerie knight," and a king, followed by his less savory improvisations as a Gyant, a Centaure, and a "storme / Raging within the waues" (39–41)—before he casts her into the dungeon where we find her at the beginning of the present canto. If Florimell is a proxy for Elizabeth, then Proteus's courtship of her in these various guises, in what might be considered an antimasque to the river procession, is a rather satiric presentation of a woman courted and then abandoned by several suitors because no one seems to please her (or her Parliament)—as she predicted. Robert Dudley, later one of Spenser's patrons, was decidedly audacious at Kenilworth in presenting his own affront to Elizabeth framed within more than a fortnight of rather bad theater, ill-manners, disrespect, and contests for power, and it is not surprising that she would censor a masque of a similar theme; how much more censure and censorship might Spenser have to fear in chiding the now twenty-year-older queen about the fact that her body is long-since outdated like the "map" of Florimell's prison at the beginning of the canto.

Just as Dudley's choice of entertainments undercuts their ceremonial praise of the queen, it is hard to believe that Spenser would think that any marriage pageant at all would please Elizabeth. Even aside from the touchy issue of her own marriage, Jonathan Goldberg explains how Elizabeth, "like Venus in the text, is a figure of desire who demands that those who desire her be unsatisfied . . . and woe to the courtier who, like Ralegh (like Timias), sought favor elsewhere. Other marriages were betrayals in Elizabeth's eyes." Her court was a place where courtiers sought favor but were never allowed to touch her virginal, mystical body (Goldberg 1981, 152).[27] As in a masque populated by mythical characters, Spenser's messages are "clowdily enwrapped in Allegoricall deuises";[28] but Spenser also has the advantage of geographical distance from the subject—a distance that Richard McCoy contends enables him to "subdue an authority whose words were always 'the best part of the play' to the power of his own words" (133).[29] From the remoteness of Ireland, and through his subtle "deuices," Spenser is able to touch and to explore the possibilities that Elizabeth's body has to offer, or perhaps more accurately, is now *beyond* offering.

The failure of Elizabeth's body to produce an heir, a great concern at the time of Spenser's writing, is reflected, in part, in his incorporation of genealogies as part of the river marriage. The use of genealogy is an important ceremonial convention: in many royal progresses (including Elizabeth's own coronation progress), royal genealogy launches the pageantry. Pierre Bourdieu contends that genealogy "serve[s] the function of ordering the social world and of legitimating that order" (34). Genealogy is also an important convention of the tradition of epideictic poetry in which Spenser participates. But genealogy cuts two ways: even while glorifying one's ancestors, genealogy looks forward to anticipated progeny and in Elizabeth's case reminds her of her *lack* of "progress" in producing (or even naming) an heir. Earlier in the *Faerie Queene*, in fact, Merlin's chronology of Britomart's progeny breaks off when he gets to Elizabeth (3.3). In Elizabeth's reign England is indeed no "teeming womb of kings" (*Richard II* 2.1.51).

The itemized genealogies of the rivers shroud another, more subtle, evolutionary catalogue of cartographic projects as Spenser presents ceremonial devices and representational maps throughout the canto that trace a historical and political progression of mapping projects. Beginning with the rather crude representation of a *mappamundi* that traps Florimell, Spenser proceeds through other ancient myths of water gods to more up-to-date representations of world maps, national maps, and local maps, and thereby creates for the reader a progression of geographic and cosmographic understanding. He ranges through both time and space: from the ancient to the modern and from the cosmic to the local; his maps depict increasing specificity of place and proximity to England. This canto's curious alliance between contemporary, historic, and mythic spaces is mirrored in the variety of maps of different vintages that might be viewed together in a map room or a study. As well, Spenser is representing a geography that includes the old alongside the new that was frequently depicted in a single map, just as medieval maps, in particular, did—showing, for example, the Garden of Eden next to more contemporary sites.[30]

As Spenser begins his description, we might get the impression that he is looking at a map and describing it to the reader. More likely, though, he is looking at or remembering a series of maps such as might be found in an early modern map room or study, popular venues for displaying power through cartographic representation, located in royal residences such as Hampton Court Palace and Whitehall.[31] Spenser would have seen quite a number of such maps displayed in the chambers and galleries at Leicester House, where he spent part of his younger days.[32] He has already demonstrated his familiarity with this type of display in Book 2 of *The Faerie Queene*, by inserting what seems to be a map room, or a study that includes maps, in Alma's castle, whose walls were painted "with picturals / Of magistrates, of courtes, of tribunals, / Of commen wealthes, of states, of pollicy" (2.9.53), among other things. These "picturals" of commonwealths and of states must surely include maps.[33]

As he names the world's rivers—the Nile, the Rhone, the Ganges, the Euphrates, and others (eighteen altogether)—Spenser seems to set no limits on which rivers are included or on the level of detail he will employ in his representation: they range in size from the prominent Nile, over 4,000 miles long, to the smaller Scamander and Alpheus, both under 100 miles; in other words, there seems to be no consistency of "scale" here. Spenser tells us that these are "famous rivers," and for Spenser their "fame"—that is, their place in history or their unique and renowned qualities—is the essential measure used in designating their "place on the map." The Scamander, for example, is called "Divine" because it is "purpled yet with blood / Of Greeks and Troians, which therein did die" (20.6–7). So small a river might almost be overlooked alongside the larger "Tygris," the Ister, or the Nile on the face of some world maps, but for Spenser, the Scamander's place in history assures its place on the map. His choice of rivers from those depicted on the maps that were at his disposal, whether mental or physical, emphasizes the construction of place in memory and history that is so central to de Certeau's distinction between space and place: these *places* have been transformed into *spaces* by what has transpired there. This procession, then, represents a march of history allied with places on the map, much like the "picturals" of "commen wealthes" in Alma's castle. As Spenser describes the rivers that he brings together in this canto, he enables the reader to view a series of maps, such as might be found together in a map room, by means of ecphrasis.

In this litany of the world's rivers, the western hemisphere garners special attention as the projects of mapping (and conquering) the New World are invoked in Spenser's two rather vitriolic stanzas on Ralegh's pet projects on the "Oranochy"[34] and the Amazon. Spenser lauds the "warlike women, which so long / Can from all men so rich a kingdome hold"; in fact, he praises them with: "*Ioy* on those warlike women" (22, emphasis added)—a salutation Spenser reserves for the truly great. By naming these two rivers of South America and the mythic, but still hoped-for, discovery of the Amazon women, Spenser makes his colonialist affinity with Ralegh manifest. Both urge the conquest of the empire of Guiana, although Spenser's suggestion is a bit more subtle than Ralegh's intimation in *The Discoverie of Guiana* that Spain will get the gold (and all that its possession entails) if England doesn't act quickly:

> For whatsoever Prince shall possess it, shall be greatest, and if the king of Spain enjoy it, he will become unresistible. Her Majesty hereby shall confirm and strengthen the opinions of all nations, as touching her great and princely actions. And where the south border of *Guiana* reacheth to the Dominion and Empire of the *Amazons,* those women shall hereby hear the name of a virgin, which is not only able to defend her own territories and her neighbours, but also to invade and conquer so great Empires and so far removed. (1968, 123)

In the New World, at least, geography seems to be destiny: whoever conquers Guiana with its gold will become the greatest ruler, and the anticipated admiration such an invasion will inspire among the Amazon women, Ralegh hopes, will be "unresistible" to Queen Elizabeth. As in the "divine tobacco" reference (of 3.5.32), Spenser takes another opportunity to remind England's queen that, thus far at least, hers is "an empire nowhere" (to use Jeffrey Knapp's coinage) because of England's belatedness in the New World and its preoccupation with "trifles" like tobacco rather than gold and conquest (Knapp 134–174). In this canto, then, Spenser's positioning of New World rivers forms a strategic transition rather than what may seem at first glance to be an impassioned digression—a transition between the rivers of the world and the rivers of England, toward which Spenser will now turn his attention. In this very important transition, Spenser advances his suggestion that England's position within the contemporary world and in history is dependent on this same connection with the New World, represented here by two of its rivers.

Spenser has already introduced us to the New World earlier in the poem. Besides the divine tobacco episode of Book 3, he has given the reader a lesson in New World discovery and cartography in the Proem to Book 2, where he needles those who think his depiction of "Faerie Lond" is a

> painted forgery,
> Rather then matter of iust memory,
> Sith none, that breatheth liuing aire, does know,
> Where is that happy land of Faery
>
> (2.Proem.1)

The larger message here is that places exist even before they are put on the map, as Spenser goes on to suggest:

> But let that man with better sence aduize,
> That of the world least part to vs is red:
> And dayly how through hardy enterprize,
> Many great Regions are discouered,
> Which to late age were neuer mentioned.
> Who euer heard of th'Indian *Peru?*
> Or who in venturous vessell measured
> The *Amazons* huge riuer now found trew?
> Or fruitfullest *Virginia* who did euer vew?
>
> Yet all these were, when no man did them know;
> (2.Proem.2–3.1)

As Spenser explains the location of the land of Faery (or Faerie lond) in this Proem, he is, in fact, defending against an implied charge that he, like a traveler weaving a tale of wonder, might be making the whole thing up. The evidence he uses to defend against such a charge is that places can and have existed for years without European knowledge of them. As proof, he offers examples from the New World—a world that is only "new" to Europeans, since the "Indian *Peru*," the Amazon, and "fruitfullest Virginia" existed, as Spenser points out, long before they were so recently "discouered"—"through hardy enterprize." Using the New World as proof of something that has now been seen is curious because few Europeans had actually *seen* the Amazon, Peru, or Virginia in Spenser's day; what they might have seen, however, is a *map* of the Amazon, Peru, or Virginia, or they might have seen "picturals" or engravings by the likes of Theodore deBry of the inhabitants of these lands. Thus, Spenser seems to be addressing a new map consciousness in this Proem, along with a belief held by his reader(s) that maps reify places in a way that mere description cannot do (or cannot do alone)—by making them visual. Thus, his readers are no longer satisfied with a "type" of the traveler's tale, a description of a fantasy world that may have obtained for earlier writers and readers—they now want a map. And without such a map, they will not believe in the existence of its world any more than they are ready to believe in the existence of other worlds (in the moon and other stars) without a map. For Spenser's readers, according to this Proem, it seems that places that have not been mapped do not exist.

Spenser's suggestion for finding "Faerie lond" (2.Proem.4) draws upon a map consciousness that he infers in his readers. He suggests they follow landmarks: "certain signes here set in sundry place [by which] / He may it find."[35] He chides those who are unable to read these "signes": "ne let him then admire, / But yield his sence to be to blunt and bace, / That no'te without an hound fine footing trace." For Spenser, it seems that those who cannot conceptualize space "need not apply."

The queen, though, is the perfect reader, for not only can she conceptualize space, she can use a map, and when she looks at some of the maps in currency in Spenser's day, she is likely to behold her own coat of arms and sometimes her portrait; likewise, in antique maps, she will see the realms of her ancestors:

> And thou, O fairest Princesse vnder sky,
> In this faire mirrhour maist behold thy face,
> And thine owne realmes in lond of Faery,
> And in this antique Image thy great auncestry.
>
> (2.Proem.4)

Indeed, some maps and atlases contained her lineage as well. As an added bonus, she could see the coats of arms of the nobility to whom she was related. The signs just don't get any better than that. Spenser's "faire mirrhour" here is the poem itself as a mirror of Elizabeth's realms and of her person, but the reference, coming as it does in the midst of a discussion of discovery, points us in the direction of maps as well. And now that these lands of Peru, Virginia, and the Amazon have been "discouered," and more importantly for this discussion, put on the map for all to see, Spenser would prefer that they be claimed and ruled by England.[36]

Turning to England, Spenser's catalog of English rivers seems to be a speculation on Saxton's composite map of the counties of England (1579).[37] As Spenser considers English rivers, he imbues each one with its own unique and specific traits: the Thames is "noble," the Ouze is "weake and crooked . . . and almost blind," the Kenet is "chaulky," the Lee is "wanton," the Severne is "stately," and the Welland is "fatal."[38] Spenser guides the reader through this terrain in a systematic counter-clockwise direction, reflecting the rather methodical government-sponsored county-by-county mapping of England during Spenser's day by Christopher Saxton (1579), John Norden (1590s), and William Smith (1590–1610).[39]

Spenser then departs from England to describe Irish rivers, in no discernable order, underscoring for Elizabeth's benefit the perceived disordered and untamed nature of that country—as Spenser reiterates in *A View of the Present State of Ireland*—adumbrating a contemporary map of Ireland (likewise the subject of much mapping in the late sixteenth century[40]). After mentioning many of these Irish rivers, and re-christening some, Spenser suggests that there are "many more whose name no tongue can tell" (44)—disparaging, perhaps, the strange-sounding native names for these rivers that would prompt the colonist to rename them to their own liking.[41] Finally, estate maps, the foundation of ownership claims, are suggested in the poet's mapping of the Mulla, which ran through his own estate of Kilcolman.[42]

Spenser's entire geographic pageant follows roughly the chronological order in which the various mapping projects were undertaken in Spenser's day; that is, the exploration of the New World necessitated a revamping of world maps to replace outdated single-hemisphere *mappaemundi*. This mapping activity, facilitated as it was by new scientific instruments and by greater mathematical accuracy, seems to have spurred a desire for greater knowledge of the local geography of England, and of Ireland, and for more accurate measurement and recording thereof. Mapping projects then focused increasingly on local geography as county, city, and estate maps followed.

By the late sixteenth century, a premium was placed on the scientific accuracy of maps, as new maps advertised their corrections of the mistakes of their predecessors. Consider, for example, the case of William Camden's *Britannia* published in 1594 and of Ralph Brooke's rapid riposte, "A Discoverie of certaine errours published in the much-commended *Britannia*" also published in 1594, which was followed summarily by another edition from Camden, "to which is added Mr. Camden's answers to this Book." Less polemic claims of accuracy and correction are often touted in the titles of new maps, such as Thomas Porter's "Newest and Exactest MAPP of the most Famous Citties LONDON and WESTMINSTER . . ." (1655). Seen in the light of his Proem about "Faerie lond," Spenser's progression hints at the inherent inaccuracy and replaceability of all maps—that is, their destiny in being corrected and replaced by a new generation of even more accurate maps. Like a pageant of genealogy, his historical sequence of figurative contemporary maps legitimates each new generation of maps by showing its no longer vital predecessors. In a similar way, Elizabeth will one day be replaced by a successor whom Spenser, for one, might hope will be more sympathetic to his causes and will have what he considers to be the "correct" foreign policy. Thus, as Spenser conducts the reader back to ancient and medieval maps, he invites correction of these maps, while at the same time he invites correction of the "map" of the queen, her political body, her biography (and especially her chastity), and her current colonial policies— leading into Book 5's more extensive and explicit critique of those policies. Just as new maps attempt to correct previous *mis*conceptions, Spenser seems to be using the figurative maps contained in his Marriage of the Thames and Medway to chide Elizabeth for her *lack* of conception. For finally—unlike the marriage of the Tame and Isis rivers further upstream from which the Thames was conceived—in Spenser's river marriage and in the geography of England, the Thames swallows up the Medway in its estuary as it is itself swallowed up in that greatest of rivers, the one that circles the earth in the ancient T-O map and the medieval *mappamundi:* the river Ocean. Not surprisingly, then, in Spenser's poem, as in Elizabeth's reign, the rivers enact no marriage vows and no actual marriage ceremony.[43] There is, in short, a marriage nowhere: one with no real location and no progeny—a particular concern in the mid-1590s as the succession question grew increasingly urgent, as we have seen. Absent a successor, there is also a widespread fear of Protestant England's dissolution into Catholicism upon Elizabeth's death (another concern of Book 5); the dissolution of the Thames and the Medway into the Atlantic Ocean, an ocean controlled by Spain and touching Catholic countries and their possessions, could reflect this fear.

Spenser's geography throughout the canto highlights women who have been victims of rape and whose rapeability, like Florimell's, motivates the action of the men who surround them. Proteus as well as Florimell's lovers are all—putatively, at least—inspired by her vulnerability to sexual attack.[44] As we have already seen, in this canto Proteus has adverted from his attempted courtship of Florimell, through performance and improvisation in previous cantos, to the use of more coercive measures. While the topography of Proteus's prison forecloses Florimell's lovers' attempts to rape her (in the sense of carrying her off, that is), her sexual rape by Proteus himself is implicit in the "cruelty and awe" of his thrall. Finally, Florimell's imprisonment is the very thing that eventually makes her appealing to Marinell.[45] In describing Florimell's situation, Spenser deploys a discourse of rape that was also frequently used to represent the fear of invasion of the queen's natural body (and the body politic for which her body stands); this discourse of rape also underpins the discourse of discovery and colonization, where the land frequently is figured as female and conquest is figured as defloration and rape.[46] As Spenser travels from the body politic of England to newly charted lands, he chooses to chronicle mythic women in marginal lands who are potential or actual victims of rape. In the wilds of Ireland, for example, Spenser narrates the story of Rheusa, the "Nimph" who is raped by the Gyant Blomius when she wanders out of bounds and who as a consequence bears three sons—from her tears (42).[47] In America, the Amazon's female denizens (21–22) pose the unique challenge of being the only women in Spenser's landscape who might successfully resist such an attack. Because of this anticipated resistance, the masculine prowess of the men of Britain is derided and goaded into action with "shame on you, O men, which boast your strong / And valiant hearts, in thoughts lesse hard and bold, / Yet quaile in conquest of that land of gold" (22). But this challenge is also aimed at a Queen who, as we have seen, has not pursued colonialist aims with the kind of zeal that Spenser would advocate. This challenge also creates an odd moment in the sexual politics of the poem, since Elizabeth is sometimes associated in the early modern period with the mythical Amazons as well as with Britomart, who in Book 5 does indeed subdue the Amazon queen Radigund. Moreover, like the Amazons, who "so long / Can from all men so rich a kingdome hold" (22), Elizabeth held off suitors until they lost interest or went elsewhere. In Spenser's poem, the courtiers and colonial adventurers who represent Elizabeth's interests in the New World are thus encouraged to rape both the Amazons and the New World for which the Amazons metonymically stand and, by extension, to rape *her* in the New World in the persons of the Amazons.[48]

Moving back to Spenser's home in colonial Ireland, another rape is suggested in the narrative of the "Mulla"—the river Awbeg in Ireland—which Spenser himself has renamed, tamed, and claimed as every good colonizer should (41), *and* has *"taught* to weep" (as if the tears of a subjugated woman are performative and desirable). It seems that the rape of the New World, when brought home to Ireland, necessitates this move to marriage: the claim of "Mulla mine" thus combines imperial conquest and colonial possession with marital name change.[49]

Finally, the Thames and Medway Rivers themselves, which were personified as brothers in *The Shepheardes Calender,* are here male and female—the Medway has been transsexualized to female in order to fit the pattern of river marriage. Though many scholars are daunted by this transsexualization,[50] I would suggest that Mary Sidney's succession to her brother Philip Sidney as poet, translator of the Psalms, and patron of the arts might help to explicate this very gender change. Spenser's view of Mary as successor to Philip is made clear in several of his poetic works. As Spenser laments the death of Philip in *The Ruines of Time* (1591), he dedicates the poem to Mary Sidney for her brother's sake, and also pays tribute to her within the poem:

Then I will sing, but who can better sing,
Than thine owne sister, peerles Ladie bright,
Which to thee sings with deep harts sorrowing,
Sorrowing tempered with deare delight,
That her to heare I feele my feeble spright
Robbed of sense, and ravished with joy,
O sad joy made of mourning and anoy.

(316–322)

In *Colin Clouts Come Home Againe* (1595), Spenser has more praise for Mary Sidney:

. . . but in the highest place
Urania, sister unto *Astrofell,*
In whose brave mynd as in a golden cofer,
All heavenly gifts and riches locked are:
More rich than pearles of *Ynde,* or gold of *Opher,*
And in her sex more wonderfull and rare.

(486–491)

In *Astrophel* (1595), she is the slain shepherd's sister Clorinda:

The gentlest shepheardesse that lives this day:
And most resembling both in shape and spright
Her brother deare

(212–214)

And finally, one of Spenser's dedications to *The Faerie Queene* is to Mary Sidney, "Countesse of Penbroke" [Pembroke] and concludes that Philip's presence is still felt through his sister:

His goodly image living euermore,
In the diuine resemblaunce of your face;
Which with your vertues ye embellish more,
And natiue beauty deck with heuenlie grace:
For his, and for your owne especial sake,
Vouchsafe from him this token in good worth to take.

(Hamilton, *FQ* p. 743)

With these references in mind, it seems almost inevitable, then, that Spenser would change the gender of the Medway for this canto. But perhaps because she *was* formerly male, Spenser's Medway seems stronger than other female figures in Spenser's geography.[51] Again, like Elizabeth with her suitors, the Medway is described as a "proud

Nymph [who] would for no worldly meed, / Nor no entreatie to his loue be led" (8). Although we don't ever find out when or why the Medway finally relented to the Thames' wooing, the suggestion of rape so prevalent in this canto is also possible here; but since we are now back home in England, the move to marriage is necessitated even more than with the Irish Awbeg/Mulla, and the suggestion of rape becomes occluded by the marriage pageant itself. Thus, the map of the river marriage, like the dungeon/map in which Florimell finds herself at the beginning of the canto, centers on a suggestion of rape not unlike the rape of the New World represented in this poem by the call to arms against the Amazons and the rapes that are the founding myths of some of the Irish rivers in Spenser's poetry.

Of course, the very kind of union and dissolution represented by the river marriage is what Queen Elizabeth successfully evaded and avoided in her self-contained virginity. A. Bartlett Giamatti contends that Florimell "traces for us the terrifying passage from containment and compression—from identity—to dissolution in the undifferentiated natural world—the loss of self" (121). We might also remember Amoret and Scudamour's embrace at the end of the 1590 version of Book 3, when Amoret's body, "late the prison of sad paine," becomes the "sweet lodge of loue and deare delight" as she is "ouercommen quight / Of huge affection" (3.12.45) as the lovers "melt" in pleasure:

> Had ye them seene, ye would haue surely thought,
> That they had beene that faire *Hermaphrodite*,
> Which that rich *Romane* of white marble wrought,
> And in his costly Bath causd to bee site:
> So seemd those two, as growne together quite,
> That *Britomart* halfe enuying their blesse,
> Was much empassiond in her gentle sprite,
> And to her selfe oft wisht like happinesse,
> In vaine she wisht, that fate n'ould let her possesse.
>
> (3.12.46)

This melting and growing together is only *half* envied by Britomart because she knows that, at this time, fate won't allow her this kind of union with Artegall.[52] Elizabeth, reading the poem, shares her substitute Britomart's ambivalent sentiments and eschews the kind of loss of identity figured by the "two senceles stocks in long embracement" (45).[53] Her virginal body, like the island and country for which it is an emblem in the Ditchley Portrait, reminds us of what that loss of self means in geopolitical terms. An alternate reading on a more positive note is offered by Roche, who associates all the river marriages with the idea of mutability and with myths of love in a fallen world—love in marriage—and concludes that "the pageant of the Thames and Medway is the highest expression of the joy and fullness of Spenser's conception of marriage" (183).

Andrew Hadfield suggests, to the contrary, that "Within the text of the poem any hope for a coherent and stable body politic remains at the level of prophecy, longed-for events which the mirror of the poem seeks to impose upon the queen" (193). Included in this category is the marriage of the Thames and Medway:

> which unites all the British rivers (including the Irish ones) [and] precedes the marriage of Marinell and Florimell. If Florimell is read as another manifestation of Elizabeth, then the emphasis is placed upon the need for Elizabeth to marry in order for her to unite her kingdoms and secure a succession. The fact that the marriage of the rivers takes place in the hall of Proteus, the god of change, helps to undercut the security of the union and provides a direct link to the representation of Elizabeth in the "Cantos of Mutabilitie." (193)

Hadfield concludes that this "union of the kingdoms" can be seen as either a "wish-fulfilment or an urge to action, and cannot be read as unproblematic praise of Elizabeth" (193). Hadfield's comments, not unlike my own, suggest how fanciful almost any union is at this point, concluding that the river marriage may be a call to action for another kind of union—that of Elizabeth's kingdoms.

Although some have noted the queen's absence in this canto of pageantry, I think she is present everywhere: in Florimell's imprisonment, in those Amazons who need to be put in their places and with whom she was frequently identified, in Mulla's "redemption" by the man who changed her name, in Rheusa's failure to observe the boundaries of maidenly behavior (and in the consequences of such a failure), and finally, as audience to a pageant and a poem in which she is never entirely present *or* absent.[54] Spenser's beginning and ending with the Ocean, which circles the medieval *mappamundi*, underscores the Ocean's mythical status as both the source and the destiny of all rivers and reflects the queen's analogous status as both source of the poem and dedicatee. Spenser reminds us in the Proem to Book 6:

> So from the Ocean all riuers spring,
> And tribute backe repay as to their King.
> Right so from you all goodly vertues well.

(6.Proem.7)[55]

Associating Elizabeth with the medieval conception of the Ocean as "well" of all "vertues" and source of all rivers makes *her* as much a captive of the *mappamundi* as Florimell, who trades her sacred chastity belt for the saltwater girdle of Proteus's cell and finally for Marinell's watery embrace. But as things now stand, Spain virtually owns the ocean, and rivers pay tribute not only to the king "Ocean," but also figuratively to the king of Spain.[56] That Spenser's mythopoeia causes rivers to flow *from* the ocean as well as *into* it, figures the inland penetration of rivers as metaphoric sexual congress and sometimes rape. The high and low tides of the Thames physically reinforce that kind of metaphoric penetration and signal the danger of dissolution for Elizabeth. If Spenser has his way, Elizabeth will turn this inward flow outward to become the new rapist of the New World. That she has found no one with whom she can melt like the "senceles stocks" of Amoret and Scudamour means that Spenser must rewrite the end of Book 3 and propose another means for Elizabeth to "succeed": she must literally "turn the tide."[57]

Bedecked with "a bushel of pearls" in her portraits,[58] Elizabeth's association with the sea—as Venus who was born of the sea foam, as Diana who controls the tides, and as victor over the Armada—seems to imply that her destiny, if she will only embrace it, is in sea venture. If Elizabeth ignores the New World (as older maps do), her legacy, like her family tree, will be a dead letter, or so both Spenser and Ralegh have suggested to her.[59] In medieval *mappaemundi*, England appears as an insignificant island at the lower, unprivileged margin of the map (see Figure 1.2); only by embracing the New World (by the force of rape, it seems) can that marginal "insular" status be corrected, since more modern maps that include the New World locate England in a higher, more central, and more privileged position. In this canto, Spenser uses reports of the New World as a strategic transition between figurations of world maps and depictions of native English geography to bring home the point that England's future place in the world (and on the map) depends largely on possessing that "new" world, which it must wrest from Spain through oceanic domination. If maps bring pleasure, then, as Thomas Elyot asserted, Spenser seems to gain pleasure in the power of his poetry to create lexical maps with which he can approach the queen with his political agenda—a political agenda that he might otherwise have to keep to himself. J. B. Harley has suggested that where maps are concerned, "the ideological arrows have tended to fly largely in one direction, from the powerful to the weaker in society. The social history of maps, unlike that of literature, art, or music, appears to have few genuinely popular, alternative, or subversive modes of expression. Maps are preeminently a language of power, not of protest" ("Maps," 301). Spenser's presentation seems to be the exception, though, as he releases a quiver of maps in the form of advice and dissent.

Richard Helgerson suggests that Spenser is untrue to geography in this canto because he "assembles rivers whose waters would otherwise meet only in the great oceanic annihilation of fluvial identity" (1992, 142), but this assumes that all the rivers are brought together on a single map.[60] I have shown, to the contrary, that it is more likely that Spenser's project within this canto is to present a variety of maps such as those found in a map room.[61] And by means of this pageant of maps we can see "all that within the world is contained" (to again invoke Elyot)—without leaving England.[62] We might apply Wilford's summation of the T-O map to Spenser's efforts: his map (or maps) are "symbolic, ornamental, and often beautiful [but] the geographic content [is] impoverished and usually misleading." Finally, Spenser's purpose seems to be to depict "a representation of [his] mind more than of the Earth" (Wilford 45). I have shown that Spenser exploits the semblance of the *mappamundi* translated into a prison in order to hold his queen's attention captive while he chastens her with his ideas of England's correct place in the world. Like Florimell, Elizabeth and England are imprisoned in the "old" geography of the ancient and Middle Ages, according to Spenser's river myths.

Surviving inventories of the contents of royal palaces made in 1547 and 1549 following Henry VIII's death reveal a great number of maps, many of which are known to have survived at least to James I's reign.[63] Maps were on public display in galleries such as the Privy Gallery in Whitehall, a particularly suitable location for displays of power, where courtiers and foreign visitors waited for audiences. As Barber speculates:

> For most of the time they had little else to do than look and, it was hoped, be impressed by the king's knowledge of his realms, by depictions of his victories (or those of his allies), and by examples of his wise patronage of the best astronomers, artists, and cartographers of the age. Despite its charm and artistry, then, the display was every bit as much a part of the infrastructure of state power as the law courts or the coastal forts. (I, 43)

In this episode of *The Faerie Queene*, it is the queen who has "little else to do" while reading the poem than to consider the implications of several cartographic worldviews; while Florimell is held captive in a dungeon, Elizabeth too is held captive by Spenser's pageant of cartographic description and is forced to consider how her foreign policy holds her country captive in an outdated version of the world that is not unlike that dungeon.[64]

When Spenser looks back on the river marriage in the next canto of *The Faerie Queene,* he emphasizes the kind of "endlesse worke" that he had undertaken in bringing all the sea gods to Proteus's hall:

> O What an endlesse worke haue I in hand,
> To count the seas abundant progeny,
> Whose fruitfull seede farre passeth those in land,
> And also those which wonne in th'azure sky?
> For much more eath to tell the starres on hy,
> Albe they endlesse seeme in estimation,
> Then to recount the Seas posterity;
> So fertile be the flouds in generation,
> So huge their numbers, and so numberlesse their nation.
>
> (4.12.1)

Here, as Spenser contrasts the fertility of the sea with the less fruitful land, he may be once again reminding Elizabeth, as he did in the marriage pageant itself, of her failure to produce an heir and pressing upon her the necessity of marrying herself to sea venture if she wants to produce "flouds of generation" for posterity.

CHAPTER THREE

MAPS, FIGURES, AND FIGURATIVE MAPS

FEMININE GEOGRAPHY[1]

In the last chapter I argued that Spenser associates potential colonies with the female anatomy—that is, the land becomes feminine. In this chapter, I extend that discussion, but I also show another side of this theme—here, the female body itself is treated in many respects as a territory to be mapped and conquered. I examine more specifically a number of literary representations of the mapping process to show how places on the female body, like places in the landscape, become invested with meaning. Faced with decisions regarding representation, mapmakers may inadvertently present biased and sometimes distorted views, as we saw in chapter 1; but maps are also sometimes deliberately used to distort evidence, as we shall see. I also expand on the discussion begun in chapter 2 of how the discovery and feminization of the New World is refracted in the literature of early modern England in the pervasive image of feminine land and masculine explorer/cartographer, along with the attendant imagery of sexual congress and rape that frequently characterizes the acts of colonizing, and of surveying and mapping. Here, I will demonstrate a link between the portrayal of Imogen in Shakespeare's *Cymbeline* and the iconography of Elizabeth I and show how the scene in which Jachimo invades Imogen's chamber enacts the process of mapping and represents the action of rape. I will show the consequences of the "soil of rape"[2] on Imogen and on "Britain," where female *subjectivity* is negated by a process of description that reduces Imogen to being the *subject* of study. Finally, I will show an instance in which a map is rejected as the representational mode of choice and an older tale is privileged instead.

As I have suggested in chapter 1, the "discovery" of the New World presented challenges to Christian doctrine as well as to cartography in that it necessitated shifting the center of the map to account for continents beyond the frame of the body of Christ, and beyond the Old Testament story of Noah's sons as founders of the three continents, as it was presented in the T-O Map, the Psalter, Ebstorf, and the Hereford Maps. The physical center of these maps at Jerusalem and the metaphoric center at Christ's navel were displaced in early modern world maps to the middle of the Atlantic Ocean, one frequent center of world maps to this day. Such an aqueous center makes the viewer long for a more stable foothold, and the New World with its heralded wonders entices the viewer as s/he seeks a place on which to land. Often figured as a woman to be ravished, the New World beckoned to explorers as well as to colonists. Not surprisingly, then, in the literature of travel and exploration, the language of desire is frequently used to advertise new discoveries enticingly. Columbus described the New World as "a land to be desired, and, seen, it is never to be left" (1:12). Similarly, Sir Walter Ralegh described Guiana as "a Country that hath yet her Maidenhead, never sacked, turned, nor wrought" (1968, 120). These two short excerpts reflect an abundance of sexual references in the literature of exploration where the land is frequently figured as a woman to be ravished and the pun on the word "country" to refer to women's genitals (as in Hamlet's "country matters" [3.2.116]) is commonplace. In America, colonies named for queens are dubbed "Maryland" and "Virginia"—the latter coinage emphasizing the virgin state of both queen and land (*both* of which presumably have yet their maidenheads), thus inviting faraway seductions that seem infeasible, and by now quite unlikely, at home. Tellingly, even colonies named for kings are feminized: Carolina for Charles and Georgia for George. The land thus feminized, discovery, exploration, invasion, and conquest can be figured as seduction, penetration, and rape. "America" itself is a feminized version of Amerigo Vespucci's first name. Jan Van der Straet's engraving of Amerigo Vespucci standing erectly over a recumbent, nude, and female "America" illustrates iconographically the initiation of what Michel de Certeau calls the "colonization of the body by the discourse of power" (1988, xxv-xxvii).[3] Annette Kolodny, too, aptly captures the sexual innuendo involved in this colonization in the title of her book on the American pastoral, *The Lay of the Land*.

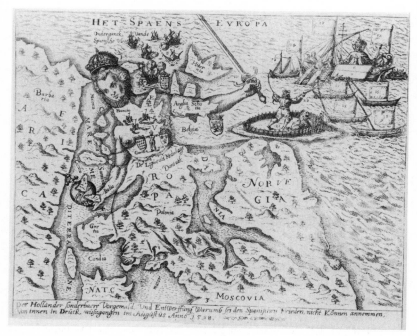

Figure 3.1 Elizabeth as Europa, An anonymous Dutch engraving (Ashmolean Museum, Oxford)

Elizabeth's body is engraved within this map of Europe. The map is oriented with west at the top and makes more sense if it is turned on its side. The queen's body covers Europe; her head is in Spain and her breast (singular because she is depicted as an Amazon) is in France; her sword-wielding arm is in England.

But topographic images of women go beyond travel and exploration narratives and iconography of the New World, showing up in both the cartography and the literature of the early modern period. In England, some of the most prominent images of woman-as-land are cartographic images of Queen Elizabeth I. A Dutch engraving of 1598 features Elizabeth as Europa (Figure 3.1); her head is Spain, her breast (singular since she is portrayed as an Amazon) is in France, her right arm is Italy, and her left sword-wielding arm is England (armed against the now-retreating papal invasion off her coast). Juxtaposed on several countries, Elizabeth's body might be seen as a symbol of power or dominion over these countries, but her body also becomes vulnerable to attack at many places—every cartographic inlet provides an orifice for invasion, or rape, as the name Europa,

PHILOSOPHORVM ΣΟΦΩΤΑΤΩ ÆSCVLAPIO SVO.

V luere cui vires & robora fana dedifti
Scribere ni vellem, na robore durior effem.
Ergo mihi (qua priuato pertingere nulli
CASE datur)tecum fatis & fatis Aftratueri eft.

SPHÆRA   CIVITATIS

ANGLIÆ · FRANCIÆ · ET · HIBERNIÆ · REGINA

Proceres   Heroes

MAIESTAS.
PRVDETIA.
FORTITVDO.
RELIGIO.
CLEMENTIA.
FACVNDIA.
VBERTAS RERVM.
IVSTITIA IMMOBILIS

ELISABETHA · D · G ·   Stellata   Confiliarii   FIDEI · DEFENSATRIX

Camera

Figure 3.2 Woodcut from John Case's *Sphæra Civitatis* (By permission of the Houghton Library, Harvard University)

Elizabeth looms above a Ptolemeic universe with attributes of her reign assigned to each sphere. Her position is reminiscent of Christ's position in the Psalter Map.

who was raped by Zeus, suggests. Similarly, in the Ditchley Portrait, as we have seen, Elizabeth the monarch overshadows and dominates the map itself and seems not just to *stand on* but to *stand in for* or to *become* Britain, the country and the island. But here again, woman-as-island is as vulnerable to attack as woman-as-continent, as we notice the tiny ships sailing under Elizabeth's skirt; this incursion, either commercial or, more menacingly, military, may remind us of Lucrece, the "late-sack'd island" at the end of Shakespeare's poem, with a sea of blood flowing around her (*The Rape of Lucrece* 1740). In *Cymbeline*, Imogen's reference to herself as "Britain" (1.6.113) and Jachimo's comment that she is "fasten'd to an empery" (1.6.120) similarly recall the Ditchley Portrait.[4]

But the iconography of Queen Elizabeth's person associated with cartographic description does not end with islands and countries. On the cosmographic scale, in a woodcut from John Case's *Sphæra Civitatas* (1588, Figure 3.2), Elizabeth's portrait looms above and embraces an outdated Ptolemaic universe over which she seems to rule by virtue of the prominence of her portrait and by the inscription of her name in the outermost sphere. But the concentric circles (representing the officers of the Court of Star Chamber) make invasion a matter of degrees, implying greater and greater intimacy within this "map" of the royal body. This image is reminiscent of the medieval *mappaemundi* discussed in chapter 1, especially the Psalter Map. In this image of Elizabeth, however, the center circle is "*Justitia Innobilis*" rather than Jerusalem or Christ's navel, but this center also seems to be located physically where the queen's genitals might be inferred beneath the cosmographic image.[5] On the microcosmic scale, the emblem of the ideal woman in the early modern period is the *hortus conclusus*, the enclosed garden—also a commonly used emblem of the queen's virginity and of England the island.[6]

Emblems of the queen are not limited to the visual realm, however, and I would argue that poems such as John Davies's *Hymnes of Astræa* (1599) represent a move from pictorial to descriptive embodiment of the monarch. Describing the queen in the formulaic manner of the acrostic highlights certain features that, taken cumulatively, might represent a linguistic map or a chorographical description of Elizabeth's person. Davies renders at least a partial delineation of a map inscribed on the monarch's name, the letters of which might be seen as the vertical axis cross-reference points or as markers of latitude on that map. For example, in "Hymne III: To the Spring," the first "E" in the recurring acrostic ELISA BETHA REGINA tells us that "Earth now is greene"; the B–E–T of "Betha" gives us "Blasts are milde and seas are calme, / Euery meadow flowes with balme, / The

Earth wears all her riches"—much like the jewel-encrusted Elizabeth of many of her portraits. In "Hymne XII: To her Picture," the poet chastises the painter for his "rude counterfeit"; a better rendering of the queen would have "each line, and each proportion right." Davies's insinuation here is of course that his poem is a better means of portraiture (or mapping), since he seems to think that he did indeed get "each line, and each proportion right" in his metrics, his diction, and his inscription of each line within the body politic represented by the name of the monarch. Words, in fact, claim superiority to pictorial representation in this description of the monarch and are powerful enough to bring this poet to something of a sexual climax in "Hymne XXIII" in his "delightfull paine," and to images of ejaculation in the middle of the final poem, "Hymne XXVI":

> Behold how my proud quill doth shed
> Eternal nectar on her head;
> The pompe of coronation
> Hath not such power her fame to spread,
> As this my admiration.

Here the pen is the mighty instrument of description that successfully competes with pictorial representation, but it is also figured sexually; for certainly a writer of Davies's skill in manipulating letters is also playing on the anagram for "Hymne"—hymen.[7]

Just as Davies's acrostics divide the monarch's name in order to describe her, John Donne's "Elegy 13: Love's Progress" divides a less regal woman's body. The poetic persona, figured as a sailor, makes a voyage on his mistress's body, describing her from the top down in geographic terms. Her brow, when smooth, is a "paradise" for the lover but when furrowed can "shipwreck" him. The lover's face is mapped out in terms of the new geography with meridians and compass points: "The nose like to the first meridian runs / Not 'twixt east and west, but 'twixt two suns" (48–49). Later, the cheek is a "rosy hemisphere" and the "swelling lips" become the "Islands Fortunate" where the lover "anchors" for "they seem all" (49–54). The poet continues with his voyage on the woman's body past the "Sestos and Abydos of her breasts" (61) and sails "towards her India, in that way / Shall her fair Atlantic navel stay" (65–66); here, the genitals are equated with exotic locales and are, of course, the real goal of this navigational inventory. Here, once again, as in the Leonardo drawing, the navel as center is displaced by the genitals. The "Islands Fortunate" and "India" (e.g., the New World) beckon the lover in the poem in the same way the New World beckons the eye in early modern world maps, as John Gillies claims that America with "its very emptiness, its nakedness perhaps (the relative absence of graphic density and verbal inscription) invites the eye to 'rove'" (62). We are well beyond the "*non plus ultra*" of the medieval maps once we are in the Atlantic of this poem. Tellingly, after leading us through an entire inventory of the female anatomy from the head down, where the lips *seem* all (and hint at the genital labia), the poet suggests that we "consider what this chase / Misspent, by thy beginning at the face" (71–72). The better route for "love's progress" would be from the bottom up—"rather set out below," the poet intones (73)—since there would be fewer distractions along the way before arriving at this "first part that comes to bed" (80). Here, then, is what the narrator sees as the center of the woman's body, or at least the center of his interest, *and* what should be at the center of his map of it. At times though, it is hard to tell whether the woman's body is being mapped or if her body is itself a map being read.[8]

Composing a poem with certain cross-references to a woman's name or "incorporating" an itemized list of geographical features into a poem may seem like an innocuous pastime; however, in another context Nancy Vickers cautions that the poetic genre of the *blazon*, which divides the woman's body while describing it, can be dangerous.[9] In poems such as *The Rape of Lucrece*, where breasts are at first "a pair of maiden worlds unconquered" (408), description leads to competitive desire, rape, and eventual death; according to Vickers, "occasion, rhetoric, and result are all informed by, and thus inscribe, a battle between men that is first figuratively and then literally fought on the fields of woman's 'celebrated' body" (96). Likewise, in the hands of a disreputable character such as Jachimo in *Cymbeline*, the activity of description in a boasting game between men destroys the reputation of a princess just as surely as the enticing tales of the New World prompted its conquest. Columbus and Ralegh figure the land as a woman to be ravished, but Donne and Shakespeare figure woman as a land or a country to be conquered; in *Cymbeline* Jachimo's penetration and mapping of Imogen's chamber represents his attempt at conquest.[10] Because Jachimo's actual failure at physical penetration of Imogen's body would effect his losing the boasting game in which he is involved, he must finally fabricate false evidence of sexual conquest in order to win the game.

Mapping the nation was a huge political project in Elizabethan England and commissioning a map was a manifestation of power, as we have seen. Because of their increasing use in defense, land ownership, travel, and colonization, the accuracy of maps was taking on greater import, and new maps, as I discussed in chapter 1, frequently included claims to accuracy surpassing their predecessors. Similarly, in the boasting game that is central to *Cymbeline,* Posthumus's estimation of his wife Imogen's purity is being questioned for its accuracy. If Posthumus is wrong about his wife, Jachimo's proposed "evidence" will correct or replace Posthumus's inaccurate or outdated measure of Imogen. Thus, Jachimo and Posthumus are engaged in a contest similar to that of Elizabethan cartographers over the accuracy of their maps, an example of which was discussed in chapter 2. In his initial encounter with Imogen, Jachimo, like these early modern mapmakers, extols his own skills and proclaims his ability to correct previous inaccuracies as he rhapsodizes on Imogen's qualities:

> What, are men mad? Hath nature given them eyes
> To see this vaulted arch and the rich crop
> Of sea and land, which can distinguish 'twixt
> The fiery orbs above, and the twinn'd stones
> Upon the number'd beach, and can we not
> Partition make with spectacles so precious
> 'Twixt fair and foul?   (1.6.32–38)

Here, Jachimo makes a pointed connection between the ability to appraise the heavens and the earth and the ability to appraise a woman's beauty and worth, as he chastises the absent Posthumus for not seeing clearly what a prize Imogen is: Posthumus and other men may have eyes with which to appreciate the "vaulted arch" of heaven and the "fiery orbs" of the stars,[11] along with the ability to comprehend the geography of "sea and land," and other features of the landscape (and the requisite skills to map the heavens and the earth), but they (and especially Posthumus) seem unable to truly appreciate the "spectacle" of female perfection with which Jachimo is now faced. But Jachimo, as mapmaker, intends to "make partition"—he will use his own cartographic vision, his figurative "spectacles" to chart "'twixt fair, and foul," to draw boundaries and to capture the essence of Imogen within those boundaries. The word "spectacle," besides referring to something upon which to gaze and to a "glass" used for aiding poor vision, was also used or implied in titles of maps such as Norden's *Speculum Britanniae,* and works about surveying such as William Cuninghams's *The Cosmographical Glass* (1559). Jachimo's vocation of mapmaker at Posthumus's behest requires that he gather certain information, certain knowledge about the spectacle before him, his "new found land" (to steal blatantly from another of Donne's elegies).

When Jachimo later enters Imogen's chamber covertly via a trunk, he postures himself as both explorer and rapist: he compares himself, in fact, to "Tarquin . . . ere he waken'd / The chastity he wounded" (2.2.12–14) as he begins to explore both the sleeping Imogen and her chamber.[12] As he describes Imogen in typical Petrarchan terms— her eyes are "enclosed lights, now canopied / Under these windows, white and azure lac'd / With blue of heaven's own tinct" (21–23)— he nearly forgets himself in his desire for a kiss, and he must remind himself of his mission:

> But my design!
> To note the chamber, I will write all down:
>   [*Takes out his tables.*]
> Such and such pictures; there the window; such
> Th' adornment of her bed; the arras, figures,
> Why, such and such; and the contents o' th' story.
>
> (2.2.23–27)

Jachimo's use of the words "design," "note," and "write," as well his locating certain features as reference points ("*there* the window"), make clear that he is formulating some sort of description or sketch of the room: he is noting the "place" of each object, or its location within the room. "Note" and "write" suggest an inventory or a written survey, but "design" points to a pictorial rendering, however rough. The "tables" or "table" that Jachimo takes out is probably a memorandum book, but the word "tables" also suggests a plane table used for surveying land, a mathematical table used for computing areas, a preexisting or preliminary "map" that he is amending (or on which he is locating all of his "such and suches"), or all three.[13] Of course, his sketchy "such and suches" may make us wonder how accurate a mapmaker Jachimo really is, but at this point Jachimo seems merely to be sketching notes for a map he might draw in greater detail later;[14] the move Jachimo makes from enraptured poetic description to more matter-of-fact note-taking or sketching, marked by his exclamatory "but my design!" (23), is underscored by the fact that the poetry suffers here, becoming halting and stilted where it was previously lush and descriptive.

Another possibility is that Jachimo is now constructing a memory system for himself, by means of which he will report the information to Posthumus. Within this memory system or memory theater, he would want to locate various parts of his presentation "here" and "there" within a room he is etching in his memory from the actual room in which he now stands. The "notes" could be seen as the *notae* that Quintillian suggests using to stimulate the memory (Carruthers 73–75); the "tables" might be seen to resemble the wax tablets of artificial memory systems. The memory "theater" in this case takes the shape of a diagram or map of the chamber. Mary Carruthers suggests that

> Successful memory schemes all acknowledge the importance of tagging material emotionally as well as schematically, making each memory as much as possible into a personal occasion by imprinting emotional associations like desire and fear, pleasure or discomfort, or the particular appearance of the source from which one in memorizing, whether oral (a teacher) or written (a manuscript page). (60)

Jachimo is certainly moved to "desire" and "pleasure" in the current scene—all the better for imprinting these images on his memory.

After Jachimo jots some notes regarding Imogen's chamber and her more accessible features, another change occurs when he suddenly realizes that the most compelling evidence of all would be "some natural notes about her body" that might enrich his "inventory" (2.2.28–30). The actual transformation of Jachimo's project suggested by this realization takes place, however, when the word "voucher" (39), usually taken to mean written proof, is used to describe Imogen's cinque-spotted mole; in a paradoxical sense, Jachimo himself becomes a "voucher" who realizes that he need no longer sketch out his proof and that the usual tools of mapmaking are immaterial to this particular assignment. Instead, the text inscribed on Imogen's body is now converted in Jachimo's mind to evidence "stronger than ever law could make" as he imagines himself delivering up to Posthumus "this secret / [that] Will force him think I have pick'd the lock and ta'en / The treasure of her honor" (40–42); the oral transmission of this one bit of evidence will carry more import, he believes, than would the written proof of the map that he began to approximate when he first took out his "tables." Moreover, once he realizes the potency of *this* voucher, he abandons his "design" of writing or drawing altogether, declaring, "No more: to what end? / Why should I write this down that's riveted, / Screw'd to my memory?" (42–44). Jachimo is certain that he can successfully "vouch" for Imogen's purported inconstancy with this evidence.[15] Not coincidentally, this moment of recognition corresponds to textual allu-

sions to both rape and sexual fulfillment, as Jachimo notices that
Imogen's book, evidently Ovid's *Metamorphoses,* is marked where another rape victim Philomela "gave up" (46), reinforcing the earlier allusion to *The Rape of Lucrece* in which Jachimo associated himself
with "our Tarquin." Also at this point, Jachimo realizes *he* has had
"enough" (46), underscoring the sexually satisfying nature of his
penetration of Imogen's room and her privacy, if not her body. Just
as Davies's poetic persona experienced "delightful paine" through his
words, Jachimo's ecstacy is dependent neither on actual penetration
of Imogen's body nor on the salacious pleasure he takes in viewing
the mole (though he certainly *does* take pleasure in viewing it); rather,
his ecstasy depends on his anticipation of his own success back in
Rome when he re-presents this evidence to Posthumus. As he reenters his trunk, Jachimo leaves Imogen's chamber with the rudiments
of his "design" but also with the sexual fodder for a traveler's tale
"screw'd to [his] memory" that he will relate with embellishments
later. It is worth noting that containers like trunks were important in
artificial memory systems; the treasury or *thesaurus* of the memory
was like a repository for things that are hidden and closed away as
well as those that are precious (see Carruthers, esp. 34–35). We
might remember that Jachimo's initial ruse for stowing the trunk in
Imogen's chamber was that it contained treasure for the emperor
(1.7.180–210). This image as a memory device is made even more
potent in Jachimo's imagining himself convincing Posthumus that he
has "pick'd the lock and ta'en / The treasure of her honor" (40–42).
Imogen's very name suggests words like "imagination," "imagine,"
or the "*imagines*" or images of artificial memory systems.

Jachimo's ocular survey or voyeuristic rape might again remind us
of the sexual exploration in Donne's "Elegy 2: To his Mistress Going
to Bed":

> Licence my roving hands, and let them go
> Behind, before, above, between, below.
> O my America, my new found land,
> My kingdom, safeliest when with one man manned,
> My mine of precious stones, my empery,
> How blessed am I in this discovering thee.
> To enter in these bonds is to be free,
> Then where my hand is set my seal shall be.
>
> (25–32)

In an ironic way, Jachimo's reentry into the trunk sets *him* free from
the temptation he might have felt, similar to Donne's poetic persona,

to let his hands rove "behind, before, above, between, [and] below."
Oddly, though, Jachimo has what he came for—he too has a "new
found land." His eyes have "discovered" her "behind," "above," and
so on, and his tongue can report her attributes with embellishments.
Being back in the trunk renders Jachimo free from being discovered
himself—which allows (or "licenses") him to tell of the "voyage upon
[Imogen]" (1.4.158) that he has figuratively made as satisfaction of
his wager with Posthumus. Jachimo's ambiguous statement that he
lodges in fear—"though this a heavenly angel, hell is here"
(2.2.49–50)—might refer to his fear before he reenters the trunk that
either Imogen's heavenly beauty or the "hell" of her vagina might
tempt him—an interesting twist on Donne's more blatantly sexual "to
enter in these bonds is to be free." This hell will be the center of any
map that Jachimo creates, just as it is the center of his erotic interest.[16]

Georgianna Ziegler affirms the nature of Jachimo's invasion as a
figurative rape when she asserts that "the woman's room signifies her
'self,' and the man's forced or stealthy entry of this room constitutes
a rape of her private space" (73). Patricia Parker, too, remarks that
"the association of a female body with a 'chamber' is finally insepara-
ble from the violation of the chamber to which her sexuality is reduced" (1987, 136). Catharine Stimpson adds that Jachimo's theft of
Imogen's "good reputation, like his penetration of her bedchamber,
is a psychic equivalent" of rape (61). As if to reinforce this figurative
rape of Imogen by means of the invasion of her private chamber, a
ridiculous penetration is attempted in the next scene by the Queen's
clownish son Cloten to win the love of Imogen (2.3). Here, Cloten
plans to "give her music a' mornings," because "they say it will penetrate" (11–12). The stage direction, "Enter musicians," follows with
further rather suggestive directions from Cloten: "Come on, tune. If
you can penetrate her with your fingering, so; we'll try with tongue
too. If none will do, let her remain; but I'll never give o'er" (14–16).
And if this music cannot penetrate, Cloten is inclined to denounce
music itself for being a vice that the "voice of unpav'd eunuch . . . can
never amend" (27–31). Several uses of the words "come," "arise,"
and references to being "up" both late and early make clear the sexual nature of this attempted penetration, but the double figuring of
castration in "unpav'd eunuch" makes it apparent that his penetration
does not (and most likely could not) "come off," a point reiterated by
Cloten to Cymbeline: "I have assail'd her with musics, but she vouch-
safes no notice" (39–40). In an odd way, too, this "penetration" reflects the move from the written to the oral (here "fingering" to
"tongue") that Jachimo enacted in Imogen's chamber.

Meanwhile, upon Jachimo's return to Rome, as he reports on the night he had "in Britain" (2.4.45), and as he amplifies his "knowledge" of Imogen (51) he replicates the move from written evidence to memorial recording initially inspired within Imogen's chamber. First, some of the details of Jachimo's rudimentary "map," for example the tapestry of "proud Cleopatra when she met her Roman" (70), are dismissed by Posthumus as possible hearsay that Jachimo could have gathered even in Rome. In order to satisfy Posthumus, Jachimo realizes that "more particulars / Must justify [his] knowledge," as he goads Posthumus with sexual innuendo in an obvious pun on carnal "knowledge" (78–79). As Jachimo describes Imogen's "chimney" and the "chimney piece" of "chaste Dian" (80–82, with likely implications of vagina, *mons veneris*, and hymen), he has set forth some compass points for a map. Looking closer at these referents, we might surmise that by establishing the chimney as situated at the "south" of the chamber as he does, Jachimo simultaneously positions the Cleopatra tapestry as north (opposite in theme to that of "chaste Dian"). But these details and directions of north and south do not satisfy Posthumus, who retorts that "This is a thing / Which you might from relation likewise reap, / Being, as it is, much spoke of" (2.4.85–7). These lines might prompt us to ask just who would be speaking so "much" about the details of Imogen's private chamber—or how many others might have "reaped" there. And the choice of the word "reaped," which figures Imogen as land to be cultivated and harvested, summons to the reader's mind the descriptions of Cleopatra by Enobarbus and the Romans: consider especially Agrippa's comment that "She made great Caesar lay his sword to bed; / He ploughed her, and she cropp'd" (*Antony and Cleopatra* 2.2.227–8). Besides the more usual sense of harvesting, the *OED* lists an "obscure" verb meaning for "reap" as "to take away by force" (i.e., to rape)—Posthumus's suggestion in this pun and anagram of "rape" seems not implausible here since both Philario and Posthumus will soon also accuse Jachimo of stealing Imogen's bracelet.[17] Likewise, Jachimo's further details—the roof of the chamber (87–88), and the andirons or the guards to the chimney (88–91, despite their metaphoric allusions)—are hardly enough to convince Posthumus that Jachimo has compromised Imogen's "honor" (91), or that he has been "on her," to invoke the common pun. Although Posthumus praises Jachimo's remembrance, merely having accurately mapped (or sketched, or described) the chamber is not enough to satisfy Posthumus and to win the wager. Even the physical evidence of the bracelet does not suffice because it might have been stolen.

It is, of course, finally with the description of the "cinque-spotted mole," the "corporal sign" (2.4.119) that Posthumus is convinced of his wife's supposed adultery. Jachimo continues:

> If you seek
> For further satisfying, under her breast
> (Worthy her pressing) lies a mole, right proud
> Of that most delicate lodging. By my life,
> I kiss'd it, and it gave me present hunger
> To feed again, though full. You do remember
> This stain upon her?
>
> (2.4.133–139)

Here, the description of the mole, Jachimo's affidavit of truth ("by my life"), his embellishment of its virtues ("worthy her pressing"), and its paradoxical ability to provoke desire even while satisfying it together become transformed into a traveler's tale of the fantastic reminiscent of Columbus's reference to the New World and of Enobarbus's description of Cleopatra's paradoxical ability to "[make] hungry / Where most she satisfies" (*Antony and Cleopatra* 2.2.236- 7).[18] Jachimo now basks in his role as fabulist as he inspires complicity in Posthumus who is, after all, a fellow traveler and—since they are both involved in the competition of the boasting game—can do nothing but concur when Jachimo poses his question, "You do remember / This stain upon her?" To answer "no" would be to admit his own dearth of "knowledge" of Imogen.[19] This question is definitely "worthy [Jachimo's] pressing" since it wins him the game. Posthumus's response, too, becomes embellished: the mole, now a "stain," in fact, becomes "as big as hell can hold" (2.4.140) as he implores Jachimo to "spare your arithmetic, never count the turns. / Once, and a million!" (142–43). Here, once again allying Imogen with Cleopatra, Posthumus multiplies Antony's epithet for Cleopatra, the Egyptian "triple-turn'd whore" depicted on Imogen's tapestry (*Antony and Cleopatra* 4.12.13), so that Imogen may as well have "turned" a million times in bed with Jachimo and other men; Posthumus's condemnation of women, too, is greatly magnified as he implicates all women and finally all things feminine in his "woman's part" speech in the next scene (2.5.1–35). The mole, a "voucher" at the site of description, becomes a "stain" only at a distance—can be exaggerated and embellished only once Jachimo is out of the chamber, and in fact, like Enobarbus, back in Rome. This exchange is a fine example of René Girard's theory, taken up by Eve Kosofsky Sedgwick, that the rivals' relationship—that is, the relationship between two men in competition for the same woman—becomes

> as intense and potent as the bond that links either of the rivals to the beloved: that the bonds of "rivalry" and "love," differently as they are experienced, are equally powerful and in many senses equivalent. For instance, Girard finds many examples in which the choice of the beloved is determined in the first place, not by the qualities of the beloved, but by the beloved's already being the choice of the person who has been chosen as a rival. (Sedgwick 21)

In the case of Jachimo and Posthumus that bond of rivalry is indeed strong, and the wager depends on it.

As a surveyor or cartographer within Imogen's chamber, Jachimo is at first restricted to accurate recording of "topographical" features of both the chamber and of Imogen, because he is there to correct Posthumus's previous "map" (or theory about his wife) in a verifiable way, but the traveler's tale, created or embellished away from the subject, offers a better medium for Jachimo's purposes because of its greater tolerance for improvisation. That this kind of improvisation is a norm in traveler's tales is evidenced in Annette Kolodny's enumeration of the range of the journeys on which such tales are based: "some planned, some already executed, some wholly imaginary, and some a confusing combination of the three" (11). Thus, the very circumstances under which the traveler's tale is written (or invented)—away from the subject locale, and often based on a thumbnail sketch such as Jachimo's—allow for improvisation, expansion, exaggeration, and complete fabrication. Jachimo may have undertaken to draw an accurate map, but he cannot be objective about a subject that threatens to carry him off in the rapture he experiences, for example, during his initial meeting with Imogen that prompts her to ask, "What, dear sir, / Thus raps you? Are you well?" (1.6.50–51). Ironically, the traveler's tale that Jachimo invents, a product of his own "rapture," turns out to be more convincing.

François Hartog outlines a rhetoric of otherness at work in travel literature such as Herodotus's *History*, the first ethnography about those living outside Greece in his day, that is useful to my discussion of Jachimo's voyage to Imogen's chamber. According to Hartog, the problem facing a writer of ethnography becomes one of translation; in other words, "how can the world being recounted be introduced in convincing fashion into the world where it is recounted?" (212). The degree to which ethnographers (and travelers) are able to convince their audiences is evaluated based on who is speaking and on the claims of experience made by the speaker. Firsthand experience, seeing with one's own eyes, or "autopsy" is the best evidence;[20] second best is hearsay evidence or experience that is reported by others; least convincing is that which is created by the traveler. Hartog uses key phrases employed by Herodotus to answer the question, "Who is speaking and to whom and how?" (211). These key phrases correspond to the levels in the evidentiary hierarchy so that "I have seen" ranks higher and indicates more direct experience and the greatest likelihood of truth; "I have heard" ranks next; and the phrases "I say" and "I write" are more likely to indicate fabrication.[21] Since vision is connected with almost certain persuasion, Jachimo succeeds in convincing Posthumus of the veracity of his tale because he is able to elevate pure fabrication of a sexual encounter into evidence of "autopsy" status. And by virtue of his having seen the mole, Jachimo is presumed to have explored the very center of Imogen's body, which he makes the center of the map he attempts to construct; he is presumed to have seen (and *done*) everything. Posthumus's objections to Jachimo's initial evidence likewise seem to rely on this evidentiary hierarchy. Hence, Jachimo's description of Imogen's chamber is rejected because the evidence it contains is low on the hierarchy: Jachimo could have *heard* about some of these details from others, because they are "much spoke of." But Jachimo eventually succeeds because he uses evidence that only he and Posthumus have *seen with their own eyes*—with the added implication here that seeing is also "doing."

To ensure his claim, Jachimo adds another element that Hartog also considers essential to the traveler's tale—marvels and curiosities (the Greek *thoma*, 230–237), like Jachimo's description of the statues that are so lifelike they are "likely to report themselves" (2.4.83) and of the cinque-spotted mole with its seemingly mystical properties.[22] By presenting *thoma*, Jachimo wins the game because he claims to have seen something of great wonder with his own eyes, and the mutual gaze of the rivals Jachimo and Posthumus endows that visual wonder with its significance as sexual evidence. By Jachimo's narration, the *places* of Imogen's chamber and of her body become *spaces* imbued with meaning, to borrow from de Certeau. The mole becomes a "space" of sexual transgression by its synecdochal equivalence.

In fact, Jachimo is never able to render an accurate map because the boasting game requires that Posthumus remain involved, and only a traveler's tale allows and accommodates such participation,

such complicity. The rules of this rivalry of description, as Jachimo
later tells Cymbeline, were established when Posthumus began

> His mistress' picture, which by his tongue being made,
> And then a mind put in't, either our brags
> Were crak'd of kitchen trulls, or his description
> Prov'd us unspeaking sots.

> (5.5.175–178)

Here, it is Posthumus's own traveler's tale, this "picture" made by *his*
tongue, requiring embellishment and complicity by his listeners
("and then a mind put in't"), along with his own boasting and "publishing," that are to blame for prompting Jachimo's action.[23]
Jachimo's tale, as he finally unravels it, shows that while he may have
been capable of drafting a true map of Imogen's terrain (Posthumus
does commend the accuracy of his remembrance, after all, in 2.4.92-
93[24]), this particular competition requires an embellished traveler's
tale instead; it requires, in effect, that a "mind [be] put in't." The interaction between these men as described by Jachimo may again
demonstrate a strong homosocial link between two men in pursuit of
the same woman. Here, it seems clear that it is Posthumus's *possession*
of a mistress that makes the others' women seem like "kitchen trulls"
and that makes Imogen so enticing.

It is significant, too, that much of the imagery in both Jachimo's
initial meeting with Imogen as well as his nocturnal penetration of
her chamber alludes to navigational matters, foreshadowing his later
move to the tale of wonder. Jachimo's mission according to Posthumus was, after all, to "make your voyage upon [Imogen]" (1.4.158).
And, as we have seen, Jachimo initially becomes so transported, sea-
sick, or "mad" by his own lust over this new terrain that Imogen has
to bring him back to reality with the query, "What, dear sir, / Thus
raps you? Are you well?" (1.6.50–51). At this initial meeting Jachimo
is able to observe only her surface qualities—her shoreline or border,
if you will—and can only speculate about what he might find with a
survey of the interior of this continent:

> All of her that is out of door most rich!
> If she be furnish'd with a mind so rare,
> She is alone th' Arabian bird.

> (1.6.14–17)

Jachimo is already enthralled with the "wonder" of Imogen at first
sight, as his comparison of her to the exotic, mythic, and wonderful

"Arabian bird"—the phoenix—makes clear. He will continue with his further penetration of Imogen's terrain via a trunk, a smaller vessel than the one in which he would have initially arrived. Like a good navigator, when he emerges from his "vessel," he notes the time of day: "The crickets sing, and man's o'erlabor'd sense / Repairs itself by rest" (2.2.11–12). He also notes what might be deemed the weather conditions: "'Tis her breathing that / Perfumes the chamber thus. The flame o' th' taper / Bows toward her" (2.2.18–20).[25] His description here is reminiscent of those New World explorers described by Annette Kolodny in *The Lay of the Land:*

> On the second of July, 1584, two English captains, Philip Amadas and Arthur Barlowe, entered the coastal waters off what is now North Carolina and enjoyed "so sweet and so strong a smell as if we had been in the midst of some delicate garden abounding with all kinds of odoriferous flowers." (10)

Exotic smells are, in fact, a common report of travelers to the New World, and Jachimo's description of this experience further associates his account with the genre of travel and exploration literature.

In another Shakespearean romance, Antonio proclaims that "Travellers ne'er did lie, / Though fools at home condemn 'em" (*The Tempest* 3.3.26), but this ironic proclamation is made after Prospero's magic has produced "*strange* SHAPES, *bringing in a banket*"; these same "shapes" then "*dance about it with gentle actions of salutations; and inviting the King, etc., to eat, they depart.*" (3.3.SD following 19). Resembling reports of hospitality by New World natives, which also included gifts of food and banquets, this same bit of theatrical artifice prompts Sebastian's avowal:

> Now I will believe
> That there are unicorns; that in Arabia
> There is one tree, the phoenix' throne, one phoenix
> At this hour reigning there.   (21–24)

In *The Tempest* Shakespeare seems to be poking fun at these castaways who suddenly believe in wonders after viewing a bit of theater, but Jachimo's own bit of improvisation has a similar, but more explosive, effect on Posthumus.

Vickers discusses this kind of description, where "on the one hand, the describer controls, possesses, and uses that matter to his own ends; and, on the other, his reader/listener is extended the privilege or pleasure of 'seeing'" (96).[26] Given the voyeuristic nature of Jachimo's sexual drive, his pleasure is indeed great and his lavish descriptions demonstrate how much he values that pleasure. The "pleasure" extended to Posthumus through Iachimo's ecphrasis, however, brings only pain. Because of the nature of the boasting contest in *Cymbeline*, Posthumus is initially in the peculiar position of being figuratively both rival mapmaker (as we have seen) and, in a sense, patron to the mapmaker Jachimo, since he commissions Jachimo's voyage by virtue of the wager. However, by the time Jachimo returns with his "map," the underlying economic paradigm seems to have shifted from patronage to purchase (as did the marketplace in maps during the Early modern period)—and Posthumus is simply not "buying" Jachimo's "map" as evidence of Imogen's alleged sexual transgression. Instead, Jachimo and Posthumus are now united in the eyewitness of the traveler's tale and are mutually enthralled by their descriptions and expansions. Back in Rome, they look from a distance upon Imogen and Britain as Other, just as Enobarbus and his Roman peers relish a communal (albeit imaginary) gaze upon Cleopatra thanks to Enobarbus's evocative report. Jachimo's descriptions, taken together, are deemed by Posthumus to show that Imogen's "bond of chastity [is] quite crack'd" (5.5.207), and enable the audience to associate her stained honor with that Egyptian "triple-turned whore." Vickers's discussion of the "buying" and "selling" of descriptions of women (97) resembles New World explorers trading in fantastic tales about the enticements of a feminized America. By becoming the subject of description, by being mapped, by becoming the "new found land"—and like that land being "turned" a million times (if only in Posthumus's and Jachimo's imaginations)—Imogen has lost the innocence implied by her more likely name "Innogen."[27] For her supposed "transgression," Posthumus orders her murder. Of course, Pisanio is unable to carry out his master's orders, but Imogen is nevertheless punished for having been the subject of description. She is exiled to the edge of the British map (in Wales), where she becomes a sort of wandering womb in the body politic (now safely unmappable because of her wandering and unreadable because of her male garb) and lives in a cave in the highly feminized, savage geography of Milford Haven for two full acts before she returns to her father's court for her final degradation.[28]

Turning for a moment to modern literature, I would like to suggest a parallel between Imogen's status at the end of *Cymbeline* and that of the main character in "Douloti the Bountiful" (1995), a story by the Indian writer Mahasweta Devi. Douloti, the daughter of a tribal bonded worker in India, becomes a bond slave prostitute who, after achieving the highest status in a house of prostitution, gradually descends to the lowest rung in that same hierarchy. Finally, having contracted venereal disease and too sick to continue in her profession, she makes a journey to a hospital for treatment, only to be directed to a more remote hospital. She decides instead to walk back to her family's village. She doesn't make it home but collapses in the night on the comforting clay of a schoolyard and dies. A rural schoolmaster, Mohan, had drawn a map of India in the clay of the schoolyard in order to teach his students nationalism in preparation for an Independence Day celebration during which he was to place the Indian flag in the middle of the map. In the morning Mohan and his students discover Douloti on the map. Gayatri Spivak explains the ambiguous ending: "The story ends with two short sentences: a rhetorical question, and a statement that is not an answer: 'What will Mohan do now? Douloti is all over India'" (128). Spivak continues:

> The word *doulot* means wealth. Thus *douloti* can be made to mean "traffic in wealth." Under the last sentence—"Douloti is all over India" [*Bharat jora hoye Douloti*]—one can hear that other sentence: *Jagat* [the globe] *jora hoye Douloti*. What will Mohan do now?—the traffic in wealth [douloti] is all over the globe. . . .
>
> Such a globalization of douloti, dissolving even the proper name, is not an overcoming of the gendered body. The persistent agendas of nationalisms and sexuality are encrypted there in the indifference of superexploitation. (128)

Similarly, by having maps and tales "encrypted" on *her* topography Imogen has experienced a metaphoric death during the play even though she has escaped the actual death that Posthumus ordered for her. In the end Imogen seems to be "fasten'd to an empery" in the manner of the dead Douloti, all over India, rather than erect and stately like the Elizabeth of the Ditchley Portrait with its own "traffic in wealth" suggested by her jewels and the tiny ships. Like Douloti, Imogen's proper name (as "Britain," the next monarch) has been dissolved; after all, she was sole heir to her father's kingdom at the beginning of the play, and it is hardly coincidental that her brothers who were kidnaped twenty years earlier and believed dead have discovered their own identities even while hers has been tainted. Interestingly, Jachimo's lewd penetration of Imogen's chamber is not enough to resurrect them—only after the *telling* of the tale do they appear. In other words, they don't appear until after she is "stained."[29]

As readers, we too have been entertained with a traveler's tale, since the boasting game at the center of *Cymbeline* is based on such a tale. Ironically, one early modern English translation of the story from Boccaccio's *Decameron* that is a likely source for Shakespeare begins by warning of the dangers of description:

> Wherein is declared that by over-liberal commending the chastity of women it falleth out (oftentimes) to be very dangerous, especially by the means of treacherers, who yet (in the end) are justly punished for their treachery.
>
>   (Second Day, Ninth Novel)

In this translation of Boccaccio those "dangers" and their consequences are incumbent upon the describer, or "treacherer"—who, in Boccaccio's story, ends up being impaled naked on a stake, anointed with honey, and devoured by wasps and mosquitoes. In another possible source, *Frederyke of Jennen,* the treacherous John of Florence, too, suffers a "shamefull death" of beheading.[30] In *Cymbeline,* by contrast, the treacherous Jachimo is forgiven—is told (by Imogen's father, no less) to "Live, / And deal with others better" (5.5.419–20), in effect, to "go and sin no more." Imogen, not so fortunate, is the real loser here, moving from her seemingly certain position as heir apparent (destined to occupy the central position in government) to a much less significant position, despite having her reputation cleared. Of course, Imogen *wants* the reunion with her husband Posthumus, though I'm not sure the audience does after his loutish behavior. Ironically, Britain, for which Imogen has been the sometimes metonymic equivalent, again pays Rome the tribute that Cymbeline initially withheld, despite having won the battle at Milford Haven. Oddly enough, in this play, winners are losers, and losers, such as Jachimo, are forgiven.

There is a useful pun on "tail" in Petruchio's retort to Kate, "What, with my tongue in your tail?" (*The Taming of the Shrew* 2.1.218), that is fitting in the story of Imogen: by having her "tail" become the subject of "tales" of travelers, especially those who would like to "make their voyage on her," Imogen has indeed ceased being Innogen; her marital chastity, as well as her chamber, by being explored, mapped, described, and published has been "cracked." Jachimo has figuratively had his tongue in her "tail" by his lewd and voyeuristic sexual fantasies, and he has had his tongue in her "tale" by describing and publishing his knowledge of her chamber and her person to others. The inexorable momentum of this descriptive competition is evident even in the final scene when Posthumus again reverses *his* opinion of Imogen and creates a sort of "counter traveler's tale" to Jachimo's, based on the same formula of the boasting game that initiated this destructive competition of description in the first place. Posthumus describes further wonders, too, as he *now* realizes that Imogen is indeed "the temple / Of Virtue" (5.5.220–221). Concluding that Jachimo's "map" was inaccurate and that his traveler's tale was false, Posthumus's former opinion or "map" of Imogen is reinstated, if a little embellished with newfound "wonders." It is perhaps not surprising that Shakespeare would prefer the tale to the map—he is himself a storyteller of the highest order. That he causes the intended map to be abandoned even in its making should not surprise us either. He is thereby able to use the process of mapping in order to make his own claim to what he sees as the more convincing means of portraiture—the tale.[31] Maps are becoming more and more accurate in Shakespeare's day, but the traveler's tale preserves the fantasy, the possibilities, and the freer reign of the imagination found in the margins and frontier regions (rather than the centers) of the older and more conjectural maps in which Shakespeare no doubt found his own inspiration for characters such as Caliban and for Othello's "Cannibals that each [other] eat, / The Anthropophagi, and men whose heads / [Do grow] beneath their shoulders" (1.3.143–5).[32] But even the "happy ending" of *Cymbeline* features Posthumus, the competing traveler, landing on Imogen's "shore" as Cymbeline proclaims, "See, / Posthumus anchors upon Imogen" (5.5.392–3). Thus, Imogen seems destined again to be treated as a "new found land," again colonized by the discourse of power that makes her subject to having all of her aspects charted and published broadly.

CHAPTER FOUR

LANDSCAPE, LABOR,
AND LEGITIMACY

After dealing with what Richard Helgerson has called "forms of nationhood," depictions of Britain as nation and island, I turn now to more local representations of the private domain of the country estate. Estate surveys, which may or may not include maps, differ from the county and national depictions of place discussed thus far in that they detail privately owned property and are conducted for the benefit of the owner of that property in assessing value and in improving land usage. And while accuracy and what I have been calling "correctness" continue to be important, in estate surveys these qualities are essential primarily as underpinnings of legitimacy, which assures clear title and ensures proper inheritance. Because of anxiety about preserving properties within lineal blood lines, the assurance of the legitimacy of heirs becomes inextricably woven into the fabric of the property. In this chapter, I examine ideologies of private property as they emerge in various prose works dealing with the techniques and the advantages of estate surveys in conjunction with the verbal "survey" reflected in a group of poems celebrating the aristocratic lifestyle of the country residence. Similar to the estate survey, "country house poems" such as Ben Jonson's "To Penshurst" (1616) use descriptions of the aristocratic house and the grounds surrounding it as a means of exalting its owner. The country house poem is a genre that has been addressed less for its topographic content than for its encomiastic elements and its social comment. But in these poems, and especially in "To Penshurst," the poet inhabits a position similar to the estate surveyor: neither poet nor surveyor is a member of the aristocracy but both are nevertheless allowed to "measure" the land and to appraise the lives of aristocrats while in their employ.

Don Wayne's study of the estate of Penshurst Place as well as of Jonson's poem celebrating it provides excellent background for my study. While Wayne hints at the topographic nature of Jonson's poem, and even provides his own schematic diagram showing how "To Penshurst" moves through the subdivisions of the property, he connects the poem neither to contemporary estate surveys nor to the many prose tracts dealing with surveying advantages and techniques. In another study, Paul Cubeta comes close to suggesting the map or survey-like qualities of "To Penshurst" when he says that

> The scene is one of fertility and productivity; the woods, shores, and hills provide pasturage for the farm animals and game for the family's dinner. But the physical details themselves form an almost geometrical pattern suggestive of a natural order and design. "The lower land," "the middle ground," "the topps . . . and Sydney's [sic] copp's" create a series of horizontal planes. (19)

But Cubeta stops short of discerning how those "horizontal planes" are similar to the information provided in a professional survey. In my comparison between "To Penshurst" and pictorial representations of property, it is also worth noting, as do Cubeta and Wayne, that the Penshurst of the poem is constructed of the basic elements of earth, air, wood, and water: "Thou ioy'st in better markes, of soyle, of ayre, / Of wood, of water: therein thou art faire" (7–8). Although both critics note that this construction marks the estate as a microcosm, neither makes the connection between this construction and the marginal pictorial representations of the four elements of earth, air, fire, and water as a major decorative convention on the face of many contemporary maps. This construction thus ties the estate to the greater macrocosmic and cosmographic, but also cartographic, realms.

The classical precedents for the seventeenth-century country house poem from the works of Martial, Horace, and Virgil have been discussed by G. R. Hibbard, Paul Cubeta, William McClung, and Don Wayne, among others. But these newer poems are not a mere borrowing or anglicization of the classical. Although the poets might import and adapt setting, occasion, and picturesque details, their social statements reflect their own perspectives within an accepted outline. Evoking these classical precedents is, however, one way in which the poets focus the reader's attention backward not only to classical antiquity but also to the not-so-long-ago past—a past of which these early modern estates still bear a trace and a longing.

Raymond Williams analyzes the economic factors associated with the country dwelling as part of a larger emerging world economic system. Williams reads the nostalgia of "To Penshurst," for example, as a search for an "unlocalised 'Old England'" (10) with an anticapitalist undercurrent. The definition of the estate of Penshurst by negatives, beginning with the first line: "Thou are *not*, PENSHURST, built to envious show" (emphasis added), prompts Williams to suggest that

> This declaration by negative and contrast, not now with city and court but with other country houses, is enough in itself to remind us that we can make no simple extension from Penshurst to a whole country civilisation. The forces of pride, greed and calculation are evidently active among landowners as well as among city merchants and courtiers. What is being celebrated is then perhaps an idea of rural society, as against the pressures of a new age; and the embodiment of this idea is the house in which Jonson has been entertained. (28)[1]

Williams considers this particular house to be an "island of Charity . . . in an otherwise harsh economy" (29). For Williams, the provident nature in which animals offer themselves up for sacrifice, which springs from a long tradition of *sponte sua* in this type of poetry, evokes for the seventeenth century the "easy consumption [that] goes before the fall," with its subsequent necessity of working for one's bread, although this labor is not specifically treated in the poem (32). In fact, in Williams's reading, labor is actually negated by his emphasis on the poem's statement that the walls were "rear'd with no mans ruine, no mans grone" (line 46).[2] Moreover, in "To Penshurst," Williams believes, labor seems to have been replaced by a provident natural order in which laborers are transformed rather conveniently into neighbors and are featured as recipients of the bounty of the lord. As a contrast to Jonson's poem, Williams enters into his discussion several poems in which the laborer and the labor itself are emphasized, including Robert Herrick's "The Hock-Cart" (1648), where the workers drink to the Lord's health but realize that they must "all goe back unto the plough . . . [and] Feed him ye must, whose food fills you" (33), and Stephen Duck's later "The Thresher's Labour," in which returning to work the morning after a feast reveals the "cheat" of the preceding day's festivities. In "To Penshurst" the fruits of the underlying labor are literally presented in the poem, but the traditional reading holds that the farmers and "clowns" bringing what has often been interpreted as "gifts" are participating in an egalitarian celebration that entails equal hospitality to all, including the poet, and hearkens back to feudal social structures. Even more recent explications that purport to read the poem in the glaring light of its economic underpinnings fail to understand that these people are not bringing gifts to the lord out of the kindness of their hearts; they are more likely paying their rent.[3]

I will analyze the nostalgia of the country house poem in terms of cartography and professional land survey, topics that Don Wayne and Paul Cubeta approach but do not explore. My analysis departs from those by Williams, Wayne, Cubeta, and others by tracing the connections between the seemingly disparate domains of cartography and survey, on the one hand, and poetry, on the other. In particular, I will investigate the shared impetus for these poetic and cartographic celebrations of place, nobility, and ownership that come into vogue at the very time in which the foundations of the aristocracy and their entitlement to property rights were beginning to crumble. I will argue that country house poems in their connection with surveying techniques combine praise and critique of the aristocracy: they not only evoke and reflect pride of ownership and glorify doctrines of virtue and good stewardship, but they also problematize gender issues. I discuss "To Penshurst" in conjunction with several works about surveying, such as John Norden's *Surveiors Dialogue* (1607, 1610, 1618), Radolph (or "Ralph") Agas's *Preparative for Plotting of Lands and Tenements for Surveigh* (1596), Valentine Leigh's *The Moste Profitable and commendable Science, of Surveiying of Landes, Tenementes, and Hereditamentes* (1578), and Aaron Rathborne's *The Surveyor in Four Bookes* (1616), in order to examine the legal issues at stake in representations of estates and to arrive at a broad sense of the ideological significance of these representations of estates from several viewpoints.

John Norden's *Surveiors Dialogue* was first published in five books in 1607, followed by a second edition enlarged to six books in 1610, and a third edition in 1618. In this popular *Dialogue,* a newly-arrived surveyor engages in highly contrived debates with a farmer, a lord, a bailey (or bailiff), and a purchaser of land, and finally convinces each member of his initially refractory audience of the particular personal benefits to be gleaned from a professional survey. The full title of the third edition gives a sense of the purpose of Norden's project:

SVRVEIORS DIALOGVE, Very profitable for all men to pervse, but especially for Gentlemen, Farmers, and Husbandmen, that shall either haue occasion, or be wil-ling to buy, hire, or sell Lands: As in the ready and perfect

Surueying of them, with the manner and Method of keeping a Court of Suruey with many necessary rules, and familiar Tables to that purpose. As also, The vse of the Manuring of some Grounds, fit as well for LORDS, as for TENNANTS. Now the third time Imprinted. And by the same Author inlarged, and a sixt Booke newly added, of a familiar conference, between a PVRCHASER, and a SVRVEYOR of Lands; of the true vse of both, being very needfull for all such as are to purchase Lands, whether it be in Fee simple, or by Lease. Diuided into sixe Bookes by I.N.

(title page, original spelling preserved, except long s's, 1618)

Norden's title page thus advertises the broad appeal of his work to various constituencies and clearly sets out his agenda for legitimizing the surveyor's work (with its rules, tables, and accurate scientific methods of measurement); he also details proper conduct of courts of survey in asserting and proving title claims and settling title disputes, and outlines methods of improving (or "manuring") the land for greater utility.

In Book I of his *Dialogue,* Norden's surveyor addresses himself to a resistant farmer, who begins by impugning the profession of the surveyor, calling it "evill," "unprofitable," "a vaine facultie," and "needlesse worke," among other things (1). The farmer enumerates various practices that evidence a lack of professionalism in the practices of surveyors, particularly the fixing of advertising bills on posts in the streets of London, from which practice he concludes,

as I have passed through London, I have seene many of their Bills fixed upon Posts in the streetes, to solicite men to affoord them some service: which argueth, that either the trade decayeth, or they are not skilfull, that beg imployment so publickely: for *vino vendibili suspensa hedera non est opus,* A good workeman needs not stand in the streetes or market place. (14)

Indeed, Radolph Agas's earlier *Preparative to the Platting of Landes and Tenements for Surveigh* (1596) includes as part of its full title a statement claiming that it is now being "published in stead of his flying papers, which cannot abide the pasting to poasts" (title page). Not surprisingly, it seems Norden's surveyor, too, "cannot abide the pasting to poasts" and couldn't agree more that this practice undermines the status of the profession, thus distancing himself from those who utilize such practices, and adding that not everyone who claims to be a surveyor is qualified to perform more than a few of the simpler tasks of the job:

> Euery one that hath but a part of the Art, nay, if he can performe some
> one, two or three parts, is not thereby to be accounted a Surueyor, as
> some Mechanicall men and Country fellowes, that can measure a
> peece of Land, and though illiterate, can account the quantity. . . .
> Some haue the skill of plotting out of ground, and can neatly delineate
> the same, and by Arithmeticke can cast vp the contents, which is a nec-
> essary poynt of a Surveyors office, but not all. (14–15)

Agas, too, complains of those not trained in surveying, whose "im-
perfections" are so "extreame" that "the people generally beginne to
doubt, whether there be any certainety, and perfection, in the operation of such instruments as are applied thereunto, yea, or not" (2).
Agas underscores his point by giving examples of unprofessional and
inaccurate surveys; one pointed example is a survey performed by a
so-called "surveyor" who was previously a plumber who had been
trained by a painter.[4]

As it turns out, though, Norden's farmer's initial objection to the
lack of professionalism in the surveyor's trade is merely a subterfuge
for his real objection to the undertaking of a land survey: that is, that
rents may thereafter be increased. According to Norden's surveyor,
however, if the farmer's concern is increased rents, he has more to
fear from the inaccurate reportage on the part of other tenants than
from the more studied project on which the surveyor is about to em-
bark. The real enemies, it seems, are those who would report falsely
against their neighbors, not the two persons in authority: here, the
Lord of the Manor and the professional surveyor. An accurate survey
will obviate this problem of inaccurate reportage.

As the farmer and the surveyor continue to quibble, the farmer insists that there is no need for a survey since the field itself is a "goodly
Map for the Lord to looke upon, better then [*sic*] a painted paper"
(15).[5] Here, the land (the permanent and natural) is contrasted with
the artificial and transitory ("painted paper"). The surveyor's response is that

> a plot rightly drawn by true information, describeth so the liuely image
> of a Mannor, and euery branch and member of the same, as the Lord
> sitting in his chayre, may see what he hath, where, and how it lyeth,
> and in whose vse and occupation euery particular is, vpon the suddaine
> view. (15)

By use of a map, Norden's Lord will be able to view, from his
"chayre,"[6] his entire property—albeit in a representational form—
and to see "euery branch and member of the same" (15). Tenants

"mislike" the survey, according Norden's Surveyor, "not that the thing it selfe offendeth them, but that by it they are often prevented or discouered of deceitfull purposes" (15). Those "deceitfull purposes," according to Norden, include actions

> to alter names, and properties, to remoue meeres, and to cast down ditches, to stock vp hedges, and to smoother [*sic*] up truth and falshood vnder such a cloake of conveniency, as before it be suspected or found out by view, it will be cleane forgotten, & none shall be able to say, This is the Land: wheras if it be plotted out, and euery parcell of free copy leased, and the rest be truly distinguished, no such trechery can be done against the Lord, but is shall be most readily reconciled. (16)

Future statements claiming "This is the Land" would not be contingent on those moveable boundaries to which Norden refers: meres, ditches, hedges. By virtue of the surveyor's visibility while perambulating the grounds and conducting the survey over several days or weeks, it becomes clear to the tenants that their whereabouts are known and that, potentially at least, they are subject to a kind of scrutiny that might prevent the types of abuses in the present that Norden names above; the recording of the survey would then obviate these deceitful practices in the future. This kind of scrutiny is also advocated in Conrad Heresbach's *Foure Bookes of Husbandry* (1577), which recommends daily rounds of supervision by the master, asserting that "the best doung for the feelde is the maisters foote."[7] A professional survey is an alternative to this kind of diurnal physical inspection, but one that nevertheless cultivates the posture of presence—thus eliminating the abuses and concealment of lands that Norden describes and ultimately rendering a more accurate or "correct" view of the land. Thus, with a professional survey, as with the other maps and depictions of place we have seen, the rubric of the "newest and exactest" is once again invoked. Not only will the surveyor uncover deceitful practices, he will also correct prior mistakes of measurements and assessment of rents; Norden insists that true measurements must be taken by a legitimate surveyor who is equipped with professional skills and competent in accepted scientific practices, rather than the rude mechanical whom Norden's surveyor impugns, and who is giving the profession a bad name—an *in*correct image, according to both Norden and Agas.[8]

Not surprisingly, the fear of abuses that the farmer voices is put to rest by the end of Book I of Norden's *Dialogue*, where the farmer offers to assist the surveyor on his perambulations and enthusiastically requests that his particular plot of land be surveyed first. By the end of Book I, indeed, readers behold a vista of bucolic bliss similar to Williams's reading of "To Penshurst," where fairness and egalitarianism appear to efface class distinctions. In fact, the farmer now *wants* boundaries drawn and might subscribe to a characterization of the walls (and other boundary markers) as described in "To Penshurst" where "none that dwell about them, wish them downe" (47).

While Norden's work provides the rationale for a professional survey, other works of the period provide the technical and legal background of tenant law and mathematical techniques that I will argue are essential to understanding the country house poem.[9] Valentine Leigh's *The Moste Profitable and commendable Science, of Surveiyng of Landes, Tenementes, and Hereditamentes* (1578) provides a detailed description of the diversity of manorial leaseholds, tenures, and customs. From the vantage point of the surveyor attempting to compute an accurate monetary valuation for an estate, Leigh enumerates the many factors to be considered—factors that include the various sources from which an owner of a manor might derive income and which a surveyor must consider in arriving at an accurate valuation. Notable among these determinants of value are the types and terms of tenure as well as the kinds of property on which rents, licenses, and patents may be granted. I want to suggest that Leigh's tract and others dealing with the laws of tenancy provide the necessary background against which to view Jonson's poem; my reading of "To Penshurst" pays particular attention to the legal and economic aspects of an estate such as Penshurst Place.

Jonson begins "To Penshurst" with a rough survey of the property based on what seems to be a perambulation, which allows the reader to picture the land and to assess the value mentally.[10] As Jonson apostrophizes Penshurst as "thou," he details the various portions of the estate for the reader:

> Thou ioy'st in better markes, of soyle, of ayre,
> Of wood, of water: therein thou art faire.
> Thou hast thy walkes for health, as well as sport:
> Thy *Mount,* to which the *Dryads* doe resort,
> Where PAN, and BACCHVS their high feasts haue made,
> Beneath the broad beech, and the chest-nut shade;
> That taller tree, which of a nut was set,
> At his great birth, where all the *Muses* met.   (7–14)

Jonson goes on to mention "the Ladies oke," the "copp's," and "the lower land, that to the riuer bends"; further perambulations take the reader through the "middle grounds," the banks, the "topps," the Medway River, the ponds, the orchard, and the garden, and finally to the walls. In his movement around the property, the poet acts as surveyor, measuring out the land in his poetry and presenting it by means of ecphrasis, so that the owner (and other readers) can view the property as well from whatever remove they might find themselves.[11] I would argue that most, though not all, of the topographic subdivisions that Jonson chooses to include in the poem represent economically viable and rentable units from which income may be used to support the nonproductive portions of the manor.[12] We might even compare this poetic catalogue to the rather detailed list of subdivisions of a typical manor with which Valentine Leigh begins the first section of his surveying manual, announcing, "And firste I will beginne to declare of all maner Rentes":

> There maie belonge to a Mannour, Landes, Tenementes, Mesuages, Burgages, Cottages, Curtillages, Toftes, Roueles, Tenementes, Mylles, Douehouses, Barnes, Stables, Gardens, Orchardes, Parkes, Warrens, Meares, Waters, Pondes, Stagnes, Fishynges, Meadowes, Cloases, Croftes, Feeldes, Pastures, Woodes, Groues, Heathes, Firzes, Moores, Marshes, Turbaries &c. (Leigh, B3^v-B4^r)

Leigh goes on to describe various terms of tenure—leasehold, freehold, copyhold—as well as various terms under which different portions of an estate might be leased. Tenants might be tenants "at will" or for a particular term; they might be tenants for life (or "freeholders"); they might hold the tenure in the land by "Anciente Demeasne" (or copyhold), by "Rente Service," or by "knight service," to name just a few of the varieties of tenure. Leigh gives brief descriptions of the various kinds of tenure, but defers to "Maister Littleton" in legal matters for more specific details of law and more explicit definitions of these tenures.[13]

After listing these types of tenure, Leigh enumerates more specifically the kinds of rent that a landowner might receive from each of these subdivisions within the manor. Owners might receive "Rent of Juistement, or Herbage" (that is rent of pastures, grazing lands, or deer parks), "Rente of Mills" (where the lord leases his mill to a tenant who can then charge other tenants for its use[14]), "Rent of Corne, or Hey" [sic] (where the rent might be "some certain number of corne, or Hay," or "where a Tith is impropriated to a Mannoure"), Rente of Fishynges, Rente of Swannes and Herneshawes [herons], Rent of Mines, Rente of Quarries, rent of "slimie or Claiye Earthe, whereof Bricke and tile is made," Rent of free Warrens ("of Conies, Hares, Herneshawes, or any other beastes or foules"), rente of "pention" or portion (for parsonages and vicarages), and Rent of workes (where tenants are bound to help the Lord in sowing, mowing, and harvest times with three or four days' work). There are also rents of licenses (to sublet land), "newe rent," and "encrease of rent" (Leigh B3v-D3v). Additionally, lords can charge "rent or yerely profite of Faires and Markettes" held on their land, rents on the profits of wood sales, rents of heaths and furses, and pannage (pasturage of swine in woods). Leigh's lengthy list includes nearly every type of charge that a lord might assess to a tenant (such as "rent charge," a kind of late fee), as well as some rights enjoyed by the lord in collecting rents and at a court of law. Returning to "Penshurst," I would like to argue, then, that far from veiling the economic exchange at the heart of this and other manors (as Williams suggests), Jonson makes economic exchange a major subject of the poem.[15] It seems to me that the impulse to read the poem absent its economic themes reflects an erroneous classification of this poem in the realm of the pastoral rather than the georgic.[16]

Also eliding the economic factors at work here, Don Wayne discusses what he calls a "chain of giving" at work at Penshurst—nature giving willingly of itself, farmers giving to the lord, and the lord giving to the poet and king (75). I would like to suggest that this interpretation, like other readings that consider the poem outside the context of the literature of tenure and survey, overlooks the actual labor that takes place at Penshurst (and other houses). Certainly, to those living within this or any other manor house, the production going on elsewhere on the property may indeed seem effortless because out of sight or imperceptible; the seemingly effortless abundance visible to them (and to the readers) and conspicuously presented in the main action of the poem, would prompt such a feeling. But even notwithstanding the possible reference to enclosure and the kind of neighbor-ruin that enclosures in the past (or elsewhere) would have caused, the reference to the walls of lines 44-46 underscores the absurdity of such an assumption of effortless abundance. These walls, we are led to believe, were "rear'd with no mans ruine, no mans grone"; but the very idea that walls made of "country stone" could possibly be built so effortlessly flies in the face of logic and should therefore make us question the purportedly effortless abundance that has preceded the introduction of this detail into the poem.[17] This irony should make us question the common reading of effortless abundance so that each of the self-sacrificing animals that steps up to the chopping block (or onto the banquet table) becomes an instantiation of the human labor that has effected its production; and for the privilege of performing such labor the tenants pay rent of all kinds, as we can see from Leigh's list.

Adding to the absurdity of the walls that were built "with no mans ruine, no mans grone" is the suggestion that no one would "wish them downe" and that they seem to offer a locale of *in*clusion rather than *ex*clusion: the small child seems to reach easily the ripe fruit dangling over the walls, and "all come in, the farmer, and the clowne" (48). Not only that, but these neighbors *seem* to be bearing gifts: "And no one empty-handed, to salute / Thy lord, and lady, though they haue no sute" (49–50). Andrew McRae summarizes how such poetic mendacity might have evolved:

> Throughout the period . . . poets were confronted by conflicting networks of influence; concerns of literary genre and tradition were jostled by moral, political and religious considerations. The resulting poetry typically suppresses considerations of social and economic process beneath predominantly celebratory images of a countryside bursting with produce and nurturing values of rural community. By the middle of the seventeenth century, poetic representations of agrarian society consistently promoted a vision of "merry England," which provided a crucial site of confluence for interests of property, social hierarchy, political stability and economic prosperity. (1996, 263)

But what looks at first like an egalitarian collection of people, a "rural community" jointly partaking in the bounty of the land, is really not that at all. Just as the descriptions of the self-sacrificing animals disguise the actual labors necessary for their production, the appearance of unmotivated gift-giving and goodwill belies the fact that this is most likely Rent Day and each subdivision of the land that was named earlier must be accounted for through payment of rent by its tenant.[18] Leigh notes several times in his manual that "rentes" are due at "ii or iiii [*sic*] feastes" during the year, and that rent can sometimes be paid, as Leigh makes clear, in "Capons, Hennes, Pepper, Cummunseede, or such like"(Ci[v]).[19]

The painting of *Rent Day* by Pieter Brueghel the Younger vividly depicts another perspective on the occasion of Rent Day (Figure 4.1[20]). Here, instead of depicting rent day as a festival, Brueghel portrays worried-looking, ragged-clad, labor-weary tenants who cower with hats in hand while an intimidating and exacting landlord scours the details of what appear to be lease documents, apparently checking to see if the various "offerings" of eggs, grapes, and poultry will

Figure 4.1 Rent Day, Pieter Brueghel the Younger (The Burghley House Collection. Photograph: Photographic Survey, Courtauld Institute of Art)

satisfy the stipulations of the leases.[21] Not being privy to the details of the leases that the lord in the painting reads, we don't know if *any* of these offerings of "rent in kind" will be acceptable; judging by the torn documents scattered on the floor, it seem as if some proffered "rents" have not been acceptable or perhaps not sufficient. Brueghel's pictorial *Rent Day,* like Jonson's poetic "To Penshurst," features an overflowing table—but Brueghel's is overflowing with money and legal documents. This landlord seems to have as little need of the proffered poultry and fruit as the lord at Penshurst might be presumed to have for the "gifts" that his "neighbors" bring to a table already overflowing with food.[22] In fact the overflowing board of Penshurst's great hall, the abundance described in the first half of the poem, and the ripe fruit hanging over the wall (suggesting a paradise within) would lead one to believe otherwise. Another, but complementary, reading of "To Penshurst" might see *laborers,* rather than tenants, presenting something like "an emblem of themselves in plum or pear"—that "emblem" is then an emblem of the *product* of their labor, of themselves *and their labor,* or of themselves *as labor*— that is, their very ability to produce. In other words, at least some of these people might be wage laborers who have only their own labor, represented here as an "emblem of themselves," to offer. While McRae contends that "Tenants and labourers are also incorporated into the controlling vision of natural abundance" and that "Jonson makes no attempt to depict the physical performance of agrarian labour" (1996, 288), I contend that having tenants and laborers present the fruits of their labors or their very ability to perform labor is just another, but more subtle way, of doing just that.

In his section on the lord's right of "Rente Secke," Valentine Leigh makes it evident that if rent is not paid, that is, "if it bee behinde unpaied, *after any feaste or daie of paiment,*" the Lord may re-enter and repossess the land (E1ᵛ, emphasis added). Thus, if in "To Penshurst," the country folk coming in supposedly with "no sute" are really attending the semiannual or quarterly "feaste" at which they are to pay their rent, their "sute" is implied in their offerings: the fruit, bread, and cheese that they bear *is* their suit, is their rent (or represents it), and their implied request is that they not be evicted the very next day—the day after the feast, the day on which the lord would be within his rights to do so, according to Leigh. In light of these customs and laws, the abundantly spontaneous production of the place must be called into question, for if the ponds are generating fish, the lands are producing crops, and so on, it is because in the portions of the estate away from the actual manor house of Penshurst, actual labor is being performed; here, the negation of that labor is metonymized in the classic tradition of *sponte sua*.

We have to wonder just why Jonson would call attention to the details to which he directs his readers. One example is the "blushing apricot and woolly peach" that hang over the wall (42). Andrew McRae reads this image as an indication of the ease of production that "requires no more labor than the grasp of a child" (1996, 288). But I would suggest that this fruit could also be read as a "type" of the forbidden fruit, only a hint of which drapes over the wall to the outside, as both a temptation and a reminder to everyone outside the walls of the apparent abundance within, an abundance that they are not usually privileged to share. I would further argue that by calling attention to the boundary of the property with this image, Jonson may also be hinting at the traditional Rogationtide ceremonies in which children are taught the boundaries of a parish or a village as a procession of citizens "beat the bounds" to exorcize any evil spirit that might lurk along or within the borders. The wall and the fruit of the poem represent both the limit of the village and the limit of the estate that must be taught to the next generation. Quentin Cooper and Paul Sullivan explain the tradition:

> Children were originally the chief boundary-beaters thrashing away with their sticks on the relevant stone, tree or other landmark which marked the edge of a town or parish. And the children were, in turn, beaten themselves, receiving a coin for their pains. Boys were pummelled with the sticks, ducked in waymarking ponds, dragged through intruding hedges, and even had to climb over buildings that straddled the boundary. This instilled in them a sense of place, with a wound for every landmark. (124)

I would argue, then, that Jonson deploys the image of the child at the boundary and the beat of his poetry to "beat the bounds" of the Penshurst estate. Although we do not get all the details of the boundaries of Penshurst from Jonson's poem, Jonson has positioned the child reaching for the fruit in a central position in the poem as a corporeal reminder of the boundary between estate and village. Other boundaries and subdivisions are implied in phrases such as "thy lower land, that to the river bends" and "the middle ground." Moreover, the "Lady's Oak," or Sidney Oak, besides being a reminder of the Sidney family tree, seems to be another reminder of the Rogation-tide ceremony; the main boundary marker in many parishes was often a tree with the title "Gospel Oak," at which spot a major sermon would be delivered. Prayers were also said along the procession of rogation celebrations, and the word itself comes from the Latin *rogare*—to beseech. Rogation Week and Rogation Sunday (the fifth Sunday after Easter) are a part of the liturgical year, climaxing on the Thursday of that week, Ascension Day. The liturgy of Rogation Sunday and Ascension Day centers on the blessing of houses and on Christ's ordaining his disciples to cast out evil spirits and drive out serpents.[23] The ecclesiastical nature of this holiday is reflected in "To Penshurst" by the Lady's passing on of Christian values and piety to the children near the end of the poem.[24]

Jonson's poem, then, alludes to both the medieval and the early modern methods of survey. While rogation celebrations help the tenants, the neighboring country folk, and the local townspeople to understand and remember boundaries, Norden claims that a professional survey will help the Lord to know the bounds of his land, to understand which parts are planted with particular crops, and to determine which portions might be put to better use. Jonson's poem performs similar work in detailing the various divisions of the land, their uses, and their products: the walkes (9), the Mount (10), the copp's (or coppice 19–20), the middle grounds (24), woods (26), fields (29), ponds (32), orchard, and garden (39). In the first forty-four lines, in particular, Jonson performs a rough survey of the house and property of Penshurst. We can discern from this survey where various parts of the property are, their relation to other parts, and what each of these parts contributes to the manor. We do not, of course, get the exact measurement that we would get from a modern or even an early modern survey; instead, we get something more akin to the medieval version of a survey, one based on memory, tradition, natural boundaries, and relationships—the kind of survey, in other words, that was made by periodically walking the margins and marking boundaries with ditches, hedges, and stones, and by teaching these boundaries to the children. The economics of the estate are important but so are the nostalgia for and reminders of earlier methods of survey, and the poem does not so much straddle the fence—or wall—between the two as it uses one as a foundation for the other. Instead of exorcising evil spirits from in and around a particular locale, a proper early modern survey exorcises the property by investigating and eliminating discrepancies of title; instead of beating the bounds to reinforce the limits of the property, a modern survey measures them accurately and mathematically. The medieval tradition depends on memory and relationships; the modern system relies on geometry and on recourse to courts of survey to settle boundary disputes. Each method, in its own way (and for its own time), enhances the legitimacy of title in the land.[25] Jonson's poetic survey operates in many ways like the medieval survey, in which boundaries are defined by features such as fields, rivers, trees, walls, and attachments between men and land and between neighbors; I would argue that this affinity helps to define the Sidney property as an "ancient pile" (5) and to distinguish between the Sidneys and more arriviste families to which the poem alludes who build of "touch" and marble in "ambitious heaps" (101). "To Penshurst" clings at least in part to the medieval methods of measurement and medieval modes of boundary marking—here an oak, there a wall, there a river—but it also hints at the newer, and now more legitimate, mathematical, geometrical, and economic measurement provided by the professional surveyor. Jonson may have personal reasons for straddling the boundary between the two methods, especially in this particular estate poem, as we shall see.

A similar case of vacillating between old and new is represented pictorially on the frontispiece of Aaron Rathborne's *Surveyor in Four Bookes* (1616). With this image Crystal Bartolovich demonstrates that the advent of new measuring tools led to a surveying process grounded in *mathematics* rather than "duty" (268). Bartolovich's reading of this image highlights how the surveyor's mathematical measurements shown in the top image supplant medieval traditions (represented by the stock figures below of a fool and a faun from emblem books), and how depiction of the surveyor's more exacting activity is flanked in the lower image, and supported in the upper, by the figures of *Arithmetica* and *Geometria*. In Bartolovich's reading, the surveyor "discovers, rather than helps to produce, spaces. The illustration works to convince us that the surveyor is not so much turning toward something new as revealing what was simply there all along, but unavailable to the 'blind' medieval surveyors" (278). According to Bartolovich, then,

> Mathematics provides a strategy for conducting difficult social and economic negotiations and offers a means by which the surveyor can simultaneously invest himself with authority and divest himself of responsibility in the shift from one way of viewing the land to another. The surveyor whose practices are justified by the supposed unity, purity, and neutrality of mathematics can claim that resistant tenants, who rely on mere memory and experience to organize their spatial relations, are fools; he may then argue that mathematics did all the work. (275)

The earlier method of surveying to which Bartolovich refers was propounded by Master Fitzherbert in his *Boke of Surveying*, the first surveying manual in English, published in 1523. In this work, entries in a book of survey record relational systems between land and tenants. For example,

> W C. holdeth a tenement of ye lorde lyeng bytwne the said tenement of R L on the Eest syde and the medowe called west medowe on the West and Southe and the sayd way that leadeth from A unto B on the north syde & conteyneth xiiii perches in brede & xvii in length and payeth. &c. suite of court & herryot. &c. (xxxviii^r)

Fitzherbert provides no measurement guidelines, however, and the one measuring unit he mentions, the "perch," is a unit that varied from manor to manor and from region to region in medieval and early modern England. Fitzherbert recommends that surveyors examine the fields on foot and interview tenants, but he does not detail how to take measurements. As a result, according to Bartolovich,

> The records he describes do not insert tenants into an abstractly mapped space so much as into a set of relationships to land and other people: the tenant holds of a certain lord (who is discouraged elsewhere in Fitzherbert's book from raising rents); his land abuts a field held by a named neighbor ("R L") and recognizable (but changeable) landmarks such as a meadow and a highway. The description is localized and, in its insistence on specific human ties, explicitly social. (278)

By the time Norden wrote his *Dialogue* in 1607 and Rathborne wrote his treatise in 1616, these kinds of records were no longer considered adequate for compiling a survey. By contrast, Norden and Rathborne emphasize the mathematical and scientific precision that they both employ. In discussing the frontispiece to Rathborne's book, Bartolovich explains that the professional surveyor pictured in the upper portion is

> so focused on his angle-measuring instrument, a tool just coming into use by some of the more technically sophisticated surveyors in England, that he seems to be completely indifferent to the two iconographic figures, a fool (in a cockscomb cap) and a faun (with pointed ears), who lie under his prominently spurred boots. (260)

Her interpretation of this frontispiece is that the surveyor, in order to produce a "view" (the most common name for a survey produced with instruments), must turn *toward* that view and turn *away* from other things—and some of the things that are "ruled out" might be the old way of doing things that seem to be represented by the fool and the faun. For the newer surveys, "a tenant's memory and experience matter less than the surveyor's measurement; information which used to be meaningful for describing the sizes of fields such as 'how many daies mowing' for a meadow, or 'how many daies plowing' for arable land, information which only a person who works the land would be likely to know," while potentially useful, is supplanted (Bartolovich 279).[26] The lived experience of the tenants and their attachment to the land are no longer of utmost importance. The surveyor's informed view is what matters, and the surveyor's view is the gaze of an outsider "opening up local spaces to the (limited) gaze of outsiders" (279), or even, I might add, to that outsider to the work of the land—the lord sitting in his "chayre."

Thus mathematics, according to Bartolovich, gives the surveyor a way to negotiate the change between the feudal manor and the early modern manor—the justification for this transition is that the "'new' spatial relations were not new at all, but had always already existed, waiting for the surveyor to reveal them" (268). The surveyor is thereby exonerated from any blame for his work or from any harm that might befall an owner, a tenant, or a neighbor as a result. Although Jonson's poem is clearly not an appraisal of monetary value, it does give us a sense of place tied to value, or a sense of the value of the place. We might intuit production values from the various ponds, woods, and orchards; we are meant, I think, to infer value in the quality of the hospitality and entertainment—especially in the seemingly equal treatment of guests, whether king, farmer, or poet; and we are certainly directed to assay the value of religious training for the children and daily devotions for the household, and to admire the value (and necessity) of legitimacy through a compliment to "thy lady's" virtue and marital chastity. Norden demonstrates how the neglect of these same factors (or of the knowledge found in a survey) weakens the estate and harms or prejudices the "heyre"—an act that "offendeth God, deceiueth the King, and defraudeth the Common-wealth." This invocation of order would please Robert Cecil, to whom Norden's work is dedicated, to say nothing of the king, the definitive reader. Norden's specific gripe regarding order is that the negligent lord offends God, "in that he is carelesse of his blessings bestowed vpon him"; he deceives the king, "in that he wilfully disableth himselfe to doe him the seruice he oweth him in body and goods"; and he defrauds the commonwealth, "in that he disableth himselfe to giue it that assistance, that his quality and place ought to afoord: and consequently, sheweth himselfe vnworthy to ouersee matters of State and Common-wealth, that is carelesse to see vnto his owne" (29).

While Norden concerns himself with the legitimacy of the estate itself, which should be not allowed to "prejudice *his* heyre" (emphasis added, where the antecedent of "his" is "estate," 29), these concerns may remind us of Jonson's encomium to the legitimacy of the Sidney heirs at Penshurst. To Norden's thinking, a tainted estate or clouded title seems just as bad as tainted or suspicious bloodlines. Of the many comparisons that can be made between Norden's *Dialogue* and Jonson's "To Penshurst," one of the most fruitful is in the arena of legitimacy—a major concern in the establishment of manors, in the proof of a professional survey's merits, and in issues of inheritance. As Norden explains, in order to have the designation of "Manor," the land must have been granted by the monarch. At Penshurst, the Sidneys' pedigree and their ownership of the estate, while relatively new, did indeed come from a king—Edward VI.[27] While Norden stresses the legitimacy of the estate itself—its origin, its untarnished title—Jonson seems more interested in the purity of the Sidney bloodlines secured by the virtue of the lady of the manor. By either practice, the property is cleared of evil spirits, taints to title, and blemishes to heredity. And this concern for bloodlines free of blemishes to inheritance would be of great personal interest to Jonson, too, who is eager to establish himself as legitimate poetic heir to Sir Philip Sidney as a "laureate" poet.[28]

Coming to the aid of the legitimacy of the estate in "To Penshurst" are the farmers' "ripe daughters" who enter the festivities offering "an emblem of themselves in plum or pear," sacrificing their own virtue it seems so that the young women of the upper classes (in this estate as well as others) might maintain theirs. The wall that seemed to obstruct or frustrate the child reaching for the fruit now invites the neighbors and their daughters in through its open gate. One boundary crossing, then—that of the wall, which seems easy at this time of festival—protects another boundary represented in the poem by the lady's fruitful chastity.[29] The seemingly impossible transaction that might initially be suggested by the farmers' sending their daughters "whom they would commend / This way to husbands" (54–55), when looked at more closely, becomes quite feasible, if not essential to the legitimacy of the family and the estate. "Husband" could of course be read in the limited sense of marital partner, but also (and more likely) as sexual partner (without benefit of marriage) and as "husbandman" or farmer, freeholder—one who tills the ground. Since tillage is a common metaphor for sexual intercourse,

the implication would seem to be that these daughters enter the estate for sexual purposes.[30] This implication becomes clearer when we
consider how the offerings these women bring underscore this reading: both plums and pears are fleshy fruit that bear some resemblance
to, and traditional association with, the female genitalia. Especially
provocative is the emblem of the plum, since the plum is often associated with the vulva—and particularly with that of younger girls before they acquire pubic hair. "Plum" is also associated with "plum
tree" representing the entire pudenda, including the legs. Finally,
plum is associated with the adjective "plump" and the verb "to
plump up"—in this context, to become pregnant.[31] The pear, too, is
fraught with sexual imagery. In our own day, we have only to consider the paintings of Georgia O'Keefe to see this association; in Jonson's day, the bawdy conversation among Romeo's hormone-riddled
friends might be considered a typical association—the "pop'rin pear"
and the medlar fruit as metaphors for genitalia (*Romeo and Juliet*
2.1.34–38). In "To Penshurst" the farmer's daughters have been reduced to one more commodity offered up for the benefit of the men
on this estate. Their contribution to the estate is their sexual availability (perhaps at very young ages), advertised in the fruit they
bring.[32] Since the chastity of the young women of an estate must be
preserved, young men of the aristocracy must go elsewhere for premarital sex. Sex with lower-class women and anal sex with men or
women were common means of preserving the technical virginity of
the daughters of aristocratic families, who can thereafter continue to
offer "an [untainted] emblem of *them*selves" in the marriage market.
Although the boundary crossing that brings the farmers' daughters
into the estate might be seen to be potentially polluting if we read
"husband" in the limited sense of marriage partner (and especially if
that husband is assumed to be the son of the lord), it becomes salu-
tary to the legitimacy of an estate when we consider the benefit of
preserving the sanctity of the lord's daughters' bodily boundaries and
their potential fruitfulness in marriage. In this light, the farmers'
daughters become one more element of the self-sacrificing natural
order of abundance in the Penshurst of Jonson's poem. Even more
importantly, they safeguard the legitimacy of successive generations
on the estate.

As we have seen, Jonson begins "To Penshurst" by saying that
Penshurst was *not* "built to envious show" (1); by means of this contrast Jonson seeks to legitimize the construction of Penshurst as a
more "correct" style than that employed in Prodigy houses of the
time, such as Longleat, Hardwick, Knole, and Burghley House—

houses that definitely *were* built for show. But even though Penshurst may not have been built of these showy materials, Jonson immediately proceeds to incorporate terms of possession that underscore the idea that the place may not be showy, but wealth is certainly important; moreover, everything here is significant by virtue of its being owned. In fact, Jonson uses forty-six possessive pronouns in just over a hundred lines: "thy friends" (21), "thy garden flowers" (39), "thy tables hoard" (71), "thy good lady" (84), "his children" (91), and so on. Most uses of "thy" (twenty-six in all) refer to the house itself, but by extension, "thy" refers to the lord as well, since anything that belongs to the manor ultimately belongs to the lord. Ownership in the seventeenth century was, of course, almost exclusively in the male domain, and even the catalogue of the lady's possessions, "her linen, her plate, and all things nigh," consists of household goods that are owned by the lord.[33] The lady herself is also the lord's property, as are the children—though paternity was sometimes somewhat problematic, according to Jonson, who cannot resist disparaging women's morals in general at the very moment that he praises this one woman in particular, remarking that knowing for certain whether one is the father of one's children is, in fact, "A fortune in this age but *rarely* known" (emphasis added, 92).[34] And we are left wondering why he brings up the subject at all, if, as he claims, it is not a problem at Penshurst.[35]

Although Jonson doesn't attempt to calculate monetary value as would be the case in a professional survey, he underscores the importance of value, ownership, and wealth by using monetary terms to describe the bounty of the land: "each *bank* doth *yield* thee conies" (25), "to *crown* thy open table, doth *provide*" (27), and "thou has thy ponds that *pay* thee *tribute* fish" (32, emphases added). Here, nature is celebrated in the language of monetary exchange. And, as we have seen, the lady's chastity and marital fidelity are also a *fortune* rarely known in this age, a phrase that emphasizes the market value of the lady—her value, indeed her "fortune," seems to be limited to her virtue since she has little else to call her own, and even her virtue is deployed by the lord in the service of the legitimacy of *his* children. Since women did not often deal in the world of commerce, this monetary wordplay attributes her "fortune" to the male (it is the *lord's* fortune finally), while hinting at the monetary exchange of prostitute and customer, as well as the marriage market.[36]

Jonson creates a panoramic effect as he surveys the orchard and garden, moving from one kind of fruit to another in both space and time:

Then hath thy orchard fruit, thy garden flowers,
　　Fresh as the ayre, and new as are the houres.
The earely cherry, with the later plum,
　　Fig, grape, and quince, each in his time doth come:
The blushing apricot, and woolly peach
　　Hang on thy walls, that euery child may reach. (39–44)

As Jonson provides us an overview of Penshurst that at times resembles a survey, what finally interests him most seems to be celebrating
the sense of natural and social order: of relationships of land to land;
tenant to lord; lord to king, guest, poet, and lady. In this emphasis
on structure as well as relationships, Jonson seems to vacillate between old and new methods of looking at property. At times, the relationships he mentions hearken back to an earlier day before the
widespread use of accurate measurement, triangulation, and mathematical calculations, but the heightened organization of the estate by
its various subdivisions in the highly structured lines of a poem hints
at the scientific methodology of the professional surveyor. Some
would argue that Jonson merely follows his classical predecessors,
and that a poem is no place for exact measurement and economic appraisal; nonetheless, Jonson does manage to celebrate prosperity,
fruitfulness, and abundance, and he adroitly manages allusions to re-
sources, architectural methods, class, and aristocratic values as he
measures out the estate of Penshurst for its owner and for the poem's
readers. Most importantly, in the end, he celebrates the lord who
makes his house a home by "dwelling" there, an action of which
James I would approve, since he fears for the welfare of the nation if
the trend continues toward noblemen abandoning their land in favor
of the fripperies that the city has to offer.[37]

　　If centers are important to maps, as I have suggested, it is interesting to note what Jonson adopts as the central tropes in his "map"
of Penshurst. It is no wonder that Jonson would center his survey
on issues of legitimacy—of the house and its construction, of the
land around it teeming with symbols of the past, of the family, and
of the poet. Given these issues, it should not be surprising that the
women of the poem have been one center of attention in my discussion. But Jonson also puts himself at the center of the action in
the poem because he wanted nothing more than to secure his own
legitimacy as poet and heir to the "family" of English poets of which
Sidney was the root.[38] Richard Helgerson suggests that Jonson cannot help "obtrud[ing] himself on his work, manifestly seeking to
make it an index of his laureate standing" (1983, 101). Ann Baines

Coiro contends that "Jonson was proudly uneasy about the service of praise he offered to those who could reward him. Yet, deeply desirous of being allowed to be enclosed within a country-house garden, Jonson publishes an edgy support of the aristocratic status quo" (370). Part of Jonson's "edginess" may be seen in his covert hints at illegitimacy and his own edginess about his legitimacy as laureate successor. Jonson's cursory glance at the nonproductive portions of the property would seem to place him more definitely in the position of surveyor—for these are the areas the surveyor would also skim; they are, by contrast, areas that the owner and his family and guests would enjoy—the "country-house garden" where he would like to be enclosed, according to Coiro. Nevertheless, by inserting (or "obtruding") himself into a poem that seems so much like a property survey, Jonson is able to take visual possession as an owner of the property would, at least momentarily—"as if thou [Penshurst], then wert mine, or I raign'd here" (74)—and he manages to put himself at the center of the hospitality in the medieval hall. He also contrives to put himself on an equal footing with the very king at whose court he sought to be the center, as he (to borrow from Norden once more) "describeth so the lively image of [this] Mannor, and euery branch and member of the same," as if *he* were "the Lord sitting in his chayre."

CHAPTER FIVE

CITYSCAPES AND CITY SCRAPES

Turning to metropolitan sites, this chapter focuses on city maps and urban representations in literature. I will argue that representations of the city either attempt to display the city's best side and thereby present a "closed" or closed-off view, or they attempt a more balanced presentation of the city that opens more of the city to the viewer's apprehension. Those representations that are "closed" display the best side of the city: they exhibit its splendor, its famous landmarks, and images of which the city government is most proud. This is the kind of representation we might find today by perusing a chamber of commerce brochure or a tourist board map. Ironically, even as these representations *open up* the city to outsiders, they *close off* the parts of the city that do not show the city in the most favorable light. Contrarily, I am calling those representations "open" that deliver to the reader or viewer an impression of the city in all of its variety, and allow her/him to perceive the low points as well as the high, the slums as well as the mansions, the jails as well as the cathedrals. I will argue that open views tend to expose metropolitan problems and to be critical of those problems while closed views tend to paper over civic vexations and deliver up a city free of discord or adversity. Representations of the city, whether pictorial or literary, will be "open" or "closed" depending on the purpose of the representation. In the visual realm, depictions known as "views" (later called panoramas) close off certain streets, while town plans or maps open up the city by allowing us to see into all the streets. Examination of various early maps of the city, such as those by Braun and Hogenberg, "views" of the city, such as those by Claes Jan Visscher and Anthony van den Wyngaerde, written surveys of the city of London, such as John Stow's *A Survey of London Written in the Year 1598,* and John Norden's description and map of London (*Speculum Britanniae, The first part,* 1593), are the mainstays of this study. Other city maps are represented, I will argue, in the poetry of the time, by the pageantry of regal coronation progresses and annual Lord Mayor's Shows, and by the genre of the city comedy. Lawrence Manley has asserted that "the city was a fact of life so essential and so complex as to require countless interpretations" (1988, 347). This chapter demonstrates the range of those interpretations.

One impression of the early modern city is detailed by Steven Mullaney:

> In a pre-modern or ceremonial city, civic pageantry and annual repetitive customs provided the vehicles with which a community could chart, in its actual topography, the limits and contradictions of its authority; such ritual passages of power served to inscribe the common places of London with cultural value and significance, making the city a legible emblem or icon of community. (viii)

Here, the city is a theater of memory whose terrain is waiting to be "inscribed," but this description seems to confer value only to those parts of the city touched by pageantry, and then only as they are related to that pageantry. In other words, this view seems too dependent on pageantry and "annual repetitive customs" in charting the city's topography.[1] This is the kind of city that royal coronation progresses give us: as the monarch moves through the city, s/he marks places of history. But this is not the only way of looking at the city, as Isabella Whitney shows by skirting the ceremonial, the formal, and the historical valences of London's streets that are represented by pageantry, and creating what might now be termed a postmodern city, with little sense of order. Her circuitous route and her vitriolic (even if satiric) declamations attest to her own estrangement from and loneliness within the city's bounds, and she seems to find her only solace "without" the limits of the city and beyond the confines of her gender. In her "Wyll and Testament" (1573), Whitney gives (or "leaves") parts of the city to various groups of people. Hers is an "open" perspective even though she is herself foreclosed from complete civic participation.

City comedies, set in London, give us another open perspective: by spreading the city out before the viewer, these dramas usually distinguish between insider and outsider, citizen and stranger, and almost always feature a hapless victim (usually a country gentleman) preyed upon by both the city and its citizens. These plays map the city by mapping trade and commerce, class, and behavior. Specific places figure prominently: Guildhall, London Bridge, Westminster, the Tower, the Thames, and other sites of civic pride, but also South-wark, Smithfield, the sewers, the parks, the brothels, and the taverns (to name only a few) that are overlooked in ceremonial pageants. Areas of trade are set out and are imputed specific qualities, certain areas are seen as more sinister than others (the haunts of rogues and prostitutes, for example), and even certain directions carry valences that an early modern audience would easily appreciate. In Jonson, Chapman, and Marston's *Eastward Ho!* (1605), for example, a con-temporary audience would easily recognize what a modern audience must glean from the notes or wait for the plot to unfold: that is, that "Eastward" on the Thames from London takes one to a real place at the bend in the river called "Cuckold's Haven"; thus, the direction east itself (in this play at least) hints at an important theme (or "ori-entation") of the play by directing the reader/viewer Eastward *to-ward* cuckoldry.[2] Similarly, certain places are deemed to be the particular domain of males: the Inns of Court in *Michaelmas Term* and the universities in *A Chaste Maid in Cheapside* are but two ex-amples. Places of female dominion are restricted to more private spaces: to the house and even specific rooms within it (for example, the birthing chamber), or to bawdy houses (an interestingly com-mercialized "private space"). Shops offer a semipublic place where the sexes mingle and where the shopkeeper's wife functions in the capacity of assumed bawd or hawker of wares for the benefit of her husband—at presumably whatever cost to her virtue seems neces-sary.[3] Certain parts of the city and its environs are connected with particular behavior; thus "a chaste maid in Cheapside" is implicitly oxymoronic since Cheapside is a place frequented by prostitutes. Is-sues of fertility, fecundity, inheritance, chastity, and legitimacy that were central to the celebration of the country house are viewed from the seedier side in city comedy to become (at various times) impo-tence, sterility, infertility, barrenness, cuckoldry, bastardy, harlotry, incontinence, and disease. For example in *Michaelmas Term*, "hops" are coupled with "harlots" to produce heirs (*Inductio* 14–29), and feasts are associated with "drabs" rather than with ersatz hospitality (62). "Virginity," we are told, "is no city trade" (1.2.43); the subtlety with which extramarital and premarital sex was broached in "To Penshurst" is completely overturned here as the Country Wench be-comes at once the object of desire, a crudely-described commodity, and an assumed "pung" (3.1.73).

Gail Kern Paster, in her excellent analysis of the role of the city in the Renaissance, points out that London was seen as teeming with the worst and lowest of elements and required "a symbolic dressing down, an apparent removal of artifice" (1985, 151). Attributing a "predatory appetite" to the city, Paster equates each character's definition of "place" with his/her own self-interest (152). In much of the literature of the city, and in city comedy in particular, the city is represented as the antithesis of the idyllic countryside of the country house poem; unlike the country house poem written to praise its owner,[4] much of the literature of the city tends to be satirical in tone and to focus, as Paster has noted, "well below the navel" (154). Interestingly though, just as Jerusalem (frequently associated with Christ's navel as well as his nativity) was the center of medieval *mappaemundi,* the church continues to be at the center of maps of London and of literature in which the city plays a role. Its function, however, is as likely to be commercial as it is religious: St. Paul's is evoked in this literature as often for the commercial bookstalls that surround it as for the worship within or the preaching outside at St. Paul's Cross. Although Paster is referring to the association of city comedies with the lower body functions (that is, to sexual and excremental functions), focusing "below the navel" in cartographic representations of the city draws our attention below, that is, *south* of, St. Paul's "nave" to the areas near the Thames, Bankside, and Southwark—places associated with the lower classes, lower occupations, and lower bodily functions. These plays regularly draw profit-making crowds in that liminal area of London's "outskirts" known as the Liberties. As Mullaney is quick to point out, the margins of the city did not constitute the end of the ideological inscription of values on the city or "the ritual creation of social topology" (viii). Rather, for him, the margins represented a place no less important to this process, a place where "the contradictions of the community—its incontinent hopes, fears, and desires—were prominently and dramatically set on stage" (viii). The Liberties are a place fraught with ambiguity: the place of the leper, the traitor, the prostitute, and the player, the place relegated to all those people and activities that do not fit into the city's social scheme. The Liberties represent all the misfits, the "other-places," what Michel Foucault has termed "heterotopias."[5]

In general, city maps emerge later than the other maps we have been examining. Not until late in the sixteenth century, in fact, were maps of cities undertaken independent of fortification or navigation maps, in which any details of the city that are included at all are secondary to the map's purpose for defense or transportation. While a few major buildings might be identified on such maps, details of streets and houses generally cannot be relied upon. Houses are usually presented as a conventionalized series with standardized frontages and roofs. Early city maps tend to be either bird's-eye view (from directly above) or pictorial representations with buildings shown in three dimensions (what today might be called a panorama[6]), although perspective is ignored. Interestingly, even as these maps evolved into scale maps and outline ground-plans (or true plans) with greater detail and accuracy, city maps continued to superimpose three-dimensional pictures of important buildings on more accurate scale maps.[7] According to J. B. Harley, city maps were drawn primarily for official use in improvement schemes, expansion, planning new towns, administration, and fortification. Their symbolic meanings, however, centered on constructions of the ideal city and on celebrating antiquity, mercantile wealth, and the power of cities (1983, 31).

The first of the major maps of London from this time period is known as the Copperplate Map (ca. 1553–1559, Figure 5.1[8]). No impressions of the map, either in whole or in part, have survived,[9] but three of the original copper engraving plates used in printing this map have been recovered. The three copper plates that have survived have been discovered quite recently—one as recently as 1998.[10] The Copperplate Map must have been quite grand in scale: cartographic historians infer from the area of the city covered by the three surviving plates that there were fifteen or twenty copper plates in all and that the dimensions of the printed map were nearly five feet by eight feet. Each of the surviving plates of this map highlights major features of the city, such as St. Paul's Cathedral, and gives other features, such as housing tracts, a sort of iconic shorthand. The underlying scheme of the Copperplate Map is considered to be a true plan (that is, an accurate map to scale), to which figures of people and other features have been added, along with bird's-eye views of buildings. The copper plates themselves are quite worn, evidencing the popularity of the map printed from them. As P. D. A. Harvey points out, this map as well as its two derivatives (discussed below) would have been "a statement—a widely accepted statement—of London's size, power, and prosperity, and a guide to its maze of streets" (*Maps* 74).

Three copies of a woodcut map of unknown vintage (though usually dated ca. 1561–1566) and commonly called the "Agas" map or simply the "Woodcut Map" have also survived.[11] This map measures just over two feet by six feet and was printed from eight wood blocks:

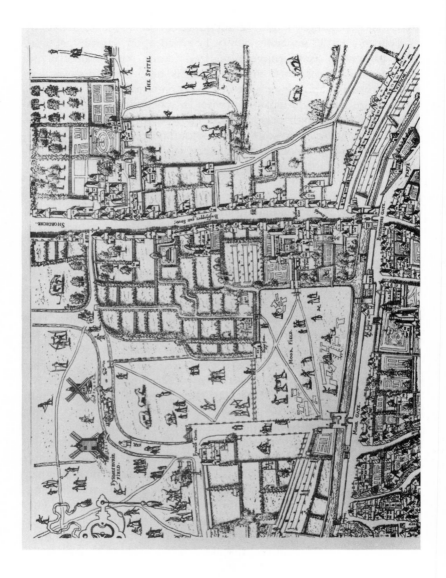

Figure 5.1 Copperplate Map of London (ca. 1553–1559) northeast section (Courtesy of the Museum of London) Londoners are busily engaged in activity around the city.

the three extant copies of the map were all printed in the seventeenth century, after some alterations were made to the original wood blocks. The Woodcut Map appears to be derived from the Copperplate Map; its inclusion of some of the same illustrative details—the same boats on the river, the dogs emerging from their kennels at the Bear Bayting Ring, among others—underscores the similarities between the two maps. The Woodcut Map does, however, seem to attempt to use perspective in the buildings on the outskirts of the city. Both the Copperplate and the Woodcut maps appear to be scale maps with pictures imposed on them of buildings and other landmarks, as well as busy denizens engaged in quotidian activities such as practicing archery, drying cloth, driving livestock, and strolling in the open spaces. Even though the map itself appears to be drawn to scale, the details of people, animals, and other objects (cannons, guns, cloth, windmills) are greatly exaggerated.

Georg Braun and Frans Hogenberg's map of London (first published in 1572) has survived in many copies (Figure 5.2).[12] This map, too, appears to be based on the Copperplate Map, though it is greatly reduced in size (with dimensions of only 13 x 19–1/8 inches, including the border). It is conceived as a true plan with a bird's-eye view of the buildings and four large figures in the foreground—in accordance with the usual style employed by Braun and Hogenberg in their monumental *Civitatis Orbis Terrarum,* or "Cities of the World."[13] These figures, according to Martin Holmes, are members of the merchant class, and this group "tells us at first sight something important about the Elizabethan Londoner: namely that he was not a courtier but a merchant" (4).[14] Georg Braun required that "towns should be drawn in such a manner that the viewer can look into all the roads and streets and see also all the buildings and open spaces" (Elliot 28). All three of the maps we have examined thus far adhere to this guideline.

Growing civic pride seems to have been the impetus behind all of these maps—pride that Braun and Hogenberg celebrated and of which they took advantage in their larger project. The civic pride captured in the London maps seems to have been contagious: a number of other English city maps of the type, with pictures superimposed on scale plans, were produced at this time by Braun and Hogenberg as well as by other mapmakers.[15] The genre of the town plan had become well established in England by the end of the sixteenth century, and these plans became the usual decorative accompaniment to county maps; thus, the famous or outstanding towns and cities of a county are given special attention by the inclusion of small city plans

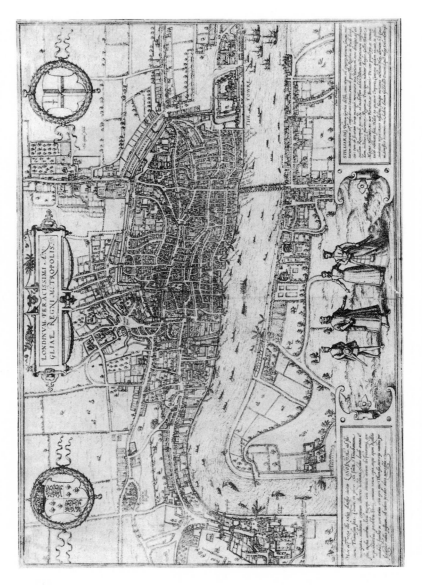

Figure 5.2 Georg Braun and Frans Hogenberg's map of London (first published in 1572; Courtesy of the Museum of London)

Notice that the church (St. Paul's) is in the center (recalling the centrality of Jerusalem on medieval maps), but our gaze has shifted to the large figures in the foreground who are members of the merchant class. This same shift is seen in the literature of the city during the Renaissance: St Paul's is evoked as often for its book stalls outside as for the worship inside.

and views in the margins of the county maps. The portions of John Norden's *Speculum Britanniae* that were completed and published, as well as John Speed's completed project *The Counties of Britain* of 1610, include city plans with nearly every county map of Britain, Ireland, Scotland, and Wales (over seventy town plans in all).[16]

John Norden's overall project is a bit different than those discussed thus far, since the map of London that he presents within his "Description of Middlesex" in his *Speculum Britanniae* (1593) is accompanied by a ten-page chorography of London (within the alphabetical index that accompanies the entire "Description of Middlesex"); this chorography includes a history of the land on which the city is situated, descriptions of the churches, parishes, and gates, and an explanation of the structure of the municipal government, among other things. Norden's prominent display of the coats of arms of each of the city's craft guilds along the side margins, in tandem with his lengthy descriptions of the various business districts, demonstrates the centrality of commerce to this city and the significance of London as the commercial hub of the country.[17] Norden's map adds to previous city maps the novel conventions of a scale and a numbered key to aid the viewer in finding specific features on the map or identifying them once found: Gray's Inn Lane, Clerkenwell, Sylver Streete, the Barbican, Newgate, and so on. His map is also considerably smaller than the others—meant for everyday use rather than display, it seems, being part of a quarto-sized book that includes chorographic descriptions. Within the same publication is Norden's map of Westminster, which likewise has a key to significant features, a scale, and a compass rose. Norden's map of the county of Middlesex includes the roads from town to town (the first map to do so), numbers and letters along the horizontal and vertical axes to aid in finding specific places on the map, and number and letter coordinates for each town, hamlet, castle, park, and other points of interest about which he writes in the chorography. Norden utilizes his own standardized symbols for market towns, parishes, castles, houses of knights, and "places where battells have bene" (*Speculum Britanniae*, map of Middlesex, between pages 14 and 15).

Norden also rendered a "Long View of London" around 1600. Not quite a map, this "view" depicts the city as if from across the Thames, preserving the sense of perspective in the buildings and thus obscuring our view of the streets. Unlike a map with a bird's-eye or "God's-eye" view, a long view is more of a panoramic view at close to eye level.[18] Interestingly, Norden has labeled the "*statio prospectiva*" of his map as St. Mary Overy's church on Bankside, and he even depicts a person on top of that tower—perhaps Norden himself—taking measurements. This "View" underscores civic pride by means of a prominent display of the Lord Mayor and twenty Aldermen in a procession (below the "view" but occupying fully one-fourth of the height of the entire work). Additionally, the view contains two inset maps (of the "linear ground plan" type) of London and Westminster, which, unlike the long view that frames them, give details of streets.[19] While Norden's "Long View" seems to straighten out the course of the Thames to a great extent, demonstrating the lack of perspective in the image, his smaller inset map of Westminster gives a more accurate picture of the bend in the river, although Norden has reoriented this map to make it more amenable to its position in the larger composition, and has thereby perhaps *dis*-oriented his reader.[20] Another curious feature of this map is its depiction of a peeling back of the landscape in the lower left corner on the "view" of *Civitas Londini* to reveal the map of Westminster beneath. The implications of this move are curious: perhaps the "view" could be seen to conceal the map (the more "open" rendering) below. And yet, Westminster would not technically be positioned on the long view quite where Norden positions this little this map.[21]

Anthony van den Wyngaerde (1543) and Claes Jan Visscher (1616) also offer up views that straighten the course of the Thames to give the viewer a fuller scene of London in their eye-level panoramas.[22] As an avatar of civic pride, the "view" serves perhaps better than the map, showing the municipal skyline and representing features of the city in a more realistic than symbolic manner, and labeling the most prominent features (Figure 5.3). Judging from their popularity, these views seem to have served their purpose well. Long views tend to be very idealized, as Ralph Hyde points out:

> In prospects and panoramas of British towns one finds no slums, no open sewers, no evidence of crime or poverty. Fine architectural landmarks predominate. Where smoke hangs over the townscape, one may be sure it has been positioned there by the artist to obscure those districts he has recorded inadequately. Town prospects are almost always gilded. (11)

For portrayals of those portions of the city that the "view" closes off from view—the inadequately represented districts, the slums, the sewers, and the parts of the city that evidence crime and poverty—we will have to look elsewhere.

John Stow's *Survey of London* is a chorographic description rather than a map or a view, but its depiction straddles the categories of open and closed views, as we shall see. As an antiquary, Stow aims to describe the city as it exists in 1598 and to recover the historical underpinnings

Figure 5.3 Claes Jan Visscher, View of London (London Bridge Section 1616) (Courtesy of the Museum of London)

Visscher's view became the basis of many subsequent panoramic views of the metropolis.

of streets, buildings, walls, conduits, and churches. After describing London's topographical situation (its "Ancient and Present Rivers, Brooks, Bourns, Pools, Wells, and Conduits of Fresh Water serving the City"[42]), Stow describes the city's man-made features, its ditches, bridges, walls, gates, and towers, and in doing so he gives the architectural history of each place, along with its human history. Stow is interested in physical dimensions and building materials, but also in the funding of building projects (to which he devotes great effort), and in the historical events associated with particular places. He also concerns himself with the customs of the city of his day, but always with an eye toward the past and the evolution of particular customs. His discussion of schools, for example, outlines how the tradition of public recitations and disputations on festival days evolved, culminating in his own experience of these events:

> For I myself, in my youth, have yearly seen, on the eve of St. Bartholomew the Apostle, the scholars of divers grammar schools repair unto the churchyard of St. Bartholomew, the priory in Smithfield, where upon a bank boarded about under a tree, some one scholar hath stepped up, and there hath opposed and answered, till he were by some better scholar overcome and put down; and then the overcomer taking the place, did like as the first; and in the end the best opposers and answerers had rewards, which I observed not but it made both good schoolmasters, and also good scholars, diligently against such times to prepare themselves for the obtaining of this garland. (101)

Generally, though, Stow places greater emphasis on the history of a place than on the type of lived experience detailed above.[23] As a rule, Stow concerns himself less with ethnography than with chorography and antiquity; today we might turn to Stow to find details of each street and the building materials for particular monuments of municipal works. Nevertheless the impulse to populate this chorography with civic denizens is apparent, associating place and people to create places of memory. We do not find as many early modern Londoners traversing Stow's streets as we do in the Copperplate Map; instead, we are likely to find the denizens of the past occupied with the events of history.

Stow describes every ward of the city in turn, its boundaries, particular streets, buildings, churches, markets, and so on. Stow walked all of the streets himself to gather his information, undertaking this project at the age of sixty. He delivers up an account to which we can still turn for detailed information about the streets and buildings that we see in the maps and views, though his account is considered to be perhaps overly nostalgic and his biases reveal themselves perhaps more in what he omits, in some cases, than in what he includes. Ian Archer, for example, points out that "Stow's silence on the matter of the lord mayor's shows is striking, given his purpose in celebrating the traditions of the City, and given the way in which they so impressed foreign visitors and other English commentators" (24); this omission seems surprising to Archer, since the city's obligations to the poor were part of these celebrations, and charity was of paramount importance to Stow (25). He also "maintains a stony silence," according to Archer, on the celebrations of the rituals of the Protestant calendar, including the celebration of Elizabeth's accession (27). Moreover, Archer considers Stow to be "unreliable" on matters of contemporary charity with his "continual harping on the hospitality of yesteryear and the laments about the 'get-rich-quick mentality' of his own time [that] convey the impression that charity had waxed cold" (27). But Archer stresses that

> Unquestionably the most significant of Stow's silences is his failure to discuss the involvement of Elizabethan Londoners in the promotion of God's word. There could be no more eloquent testimony to his lack of sympathy for evangelical Protestantism than his failure to mention the endowment by leading London merchants of parochial lectureships and the support of others by means of subscription among parishioners: by the time of the publication of his *Survey* thirty-five parishes supported lectureships. The remarkable progress made in the creation of an effective preaching ministry went unpraised by Stow. (28–29)

Archer adds that Stow's religious sympathies made him vulnerable to suspicion in his own day, reporting that a raid on Stow's study disclosed several "old fantastical Popish books" though he generally was "able to accommodate himself to the new regime in religion" (29).

Lawrence Manley contends that Stow creates a sort of "civic religion" based on ceremony, suggesting that

> Stow's *Survey* provides not only a picture of the changing metropolis but a striking example of the paradoxes of the citizen mentality, in which adherence to tradition reinforced a historic drift that transformed living civic norms into the facts of historical anthropology and Stow's civic religion into history. ("Sites" 36)

Manley also addresses Stow's relative silence on commercial theatres, his mentioning only the recent erection of "certain public places" for performances, and omitting from the 1603 edition even the names

of The Theatre and The Curtain (included in the 1598 edition).
Manley argues that

> In Stow's silence may be read an endorsement of the anti-theatrical
> bias of the London government, whose persistent attempts to restrict
> playing stemmed from many of the same concerns sounded in the
> pages of the *Survey*—including a sense of weakening control over the
> suburbs, with their illicit mingling of a shadowy variety of unenfran-
> chised strangers, aliens and nonconformists. (50)

Finally, Manley concludes that "for Stow, much of the piety that
might once have been reserved for religious veneration has now become invested in a civic religion based on veneration for the past"—
adding, interestingly, that Stow's perambulations of the ward
boundaries adapts the practice of the Rogationtide ceremony (53).
Thus, while we might at first believe that Stow presents an open view
of the City of London, it turns out that his biases make the visual
counterpart more like a closed view.

Constructing an accurate map of the city requires that the mapmaker (or someone in his employ) walk the streets in order to chart
them. Similarly, a written survey such as Stow's *Survey of London* requires the same careful attention to detail. While sometimes specify-
ing (or labeling) only one viewpoint, the long views of the city at
which we have recently been looking usually have several viewpoints,
and the details depicted most likely necessitated the closer inspection
and the rendering of multiple sketches that would have required the
artist to traverse the streets.[24] Having examined the perspective of
London presented in several maps and views that serve as manifestations of civic pride, I turn now to two very different portrayals of
London that take into account two divergent female experiences and
perspectives as they traverse London's streets.

In January 1559,[25] Queen Elizabeth made a progress through the
city of London en route to her coronation: she was welcomed with
cheers, pageants, gifts, and music. In what has been described as the
first piece of propaganda of the Elizabethan age, that progress is
chronicled in a publication, attributed to Richard Mulcaster, that appeared nine days later and was so popular that it had to be reprinted
shortly thereafter.[26] Fifteen years later Isabella Whitney, a woman lack-
ing social rank and financial wherewithal, traversed the streets of London in a quite different manner, through the lines of a poem in which
she "wills" parts of the city to various groups of people, as she
"fayneth as she would die" and retreats from London in defeat and

poverty. She memorializes the streets and makes the reader remember both the streets and the poet by the associations she has created. Although differences in rank and notoriety could account in large part for the difference in the way these two women were received by London, and the difference in genre (encomiastic propaganda versus mock will) would also have its effect, I want to look at how each of these works depicts the streets with which we have already become familiar through our investigation of contemporary maps to argue that one presentation closes off the city to our scrutiny while the other opens it up to critique, showing the reader the unsavory back streets.

Elizabeth's pageant is carefully planned and deliberately historical. According to J. E. Neale she and the citizens of London seemed to be "courting each other with delightful ardour" (7)—or so it would appear from Mulcaster's account. It seems that the coronation pageants were performed not only for the queen's benefit but for the citizens' benefit as well. The props for the pageants, though specifically made for Elizabeth's coronation, create an aura of permanence with arches, stages, scenes/backdrops, and tablets that remain behind bearing the recitations that have been delivered. Not only were arches, stages, and props erected, even the buildings and streets were dressed "to the nines" in streamers, draperies, and silken cloth, as described in the tract:

> riche hangynges, aswell of Tapistrie, Arras, clothes of golde, siluer, veluet, damaske, Sattyn, and other silkes plentifully hanged all the way as the Quenes highnes passed from the Towre through the citie. Out at the windowes & penthouses of euerie house, did hang a number of ryche and costlye banners and streamers tyll her grace came to the upper ende of Cheape. (Osborn 45 [C3^r])[27]

But this sort of "dressing up" seems to imply that London in and of itself is deficient, *must* be dressed up, covered up, hidden. The representation of London in Elizabeth's progress resembles that of a "view" where, we will remember, "one finds no slums, no open sewers, no evidence of crime or poverty," where "fine architectural landmarks predominate," and where the landscape is "almost always gilded" (Hyde 11).[28]

Not only does this pageant in which London is turned out in its finest conceal the back streets (or at least attempt to conceal them), but in its effort to show the "courting" to which Neale refers, the printed account also masks the source of financial support that underlies this particular spectacle. According to Susan Frye, the pageant was paid for by the individual members of the twelve principal guilds of the city, but sponsored overwhelmingly by the Court of Alderman. Most of these Alderman were Merchant Adventurers who depended on foreign trade of cloth to garner higher profits than were available domestically. They could not depend on support from Parliament for these endeavors and sought instead to bypass Parliament and to get a royal license in order to exceed parliamentary export limits (52–53). As Frye summarizes:

> The economic relation joining Elizabeth's interests with the Adventurers' is simply stated: Elizabeth's solvency and the aldermen's expectation of profits abroad rested on their mutual need to bypass Parliament to secure profits from the foreign sale of cloth. (52)

Part of what is foreclosed from view in *The Quene's Maiesty's Passage*, then, is the kind of transaction described by Frye, since the Merchant Adventurers also sponsored the publication of the account of the *Passage*. Not surprisingly, when the tract refers to the "city," according to Frye, "it is conflating the citizens' interests with those of the their rulers, the Court of Aldermen, by presenting the elites' opinions as being held by the entire city" (40). Lawrence Manley comments on the long-standing tradition of implied contracts, such as that established between the queen and the Aldermen:

> The City had presented a gift to Henry VI in Westminster on the day after his entry, but Anne Boleyn's entry established what thereafter became another unbroken custom—the insertion into the climactic phase of the entry of symbolic gift-giving by the City's own representatives. The offering of gifts was then followed in a ceremony in which a sword or sceptre was passed from the royal entrant to the mayor of London, who then preceded the entrant for the remainder of the procession. (*Literature and Culture* 247)

Wendy Wall, too, considers the account of Elizabeth's progress to be a bit of "state mythology," arguing that this bit of drama in which Elizabeth plays a part is highly mediated by the written account:

> By emphasizing Elizabeth's command over movement, Mulcaster weakens the challenge posed by the restructuring of royal display—the image presented of monarch and public as equal partners in an active courtship. Mulcaster instead depicts these interruptions as moments when the queen voluntarily stops for interaction with her beloved people, an arrest that reveals her truly aristocratic, "most gentle," and charitable nature. (1993, 121)

But apparently this approach was successful. Subsequent accounts borrow from Mulcaster to the point that "rather than the scenes along the route, the queen's responses became fetishized texts in numerous history books. *The Sayings of Queen Elizabeth* is an entire book devoted to reproducing these sacrosanct utterances" (122).

Also glossed over, reinterpreted, or "contained" in the Mulcaster account is an incident that occurred at "the nether ende of Cornehill," where one of the knights near the queen

> had espyed an auncient citizen, which wepte, and turned his head backe, and therwith said this gentleman, yonder is an Alderman (for so he tearmed hym) which wepeth & turneth his face backeward. Now may it be interpreted that he so doth, for sorowe, or for gladnes. The quenes maiestie hearde him, and said, I warrant you it is for gladnes. A gracious interpretation of a noble courage, which wold turne the doutefull to the best. And yet it was well known that as her grace did confirme thesame, the parties cheare was moued for very pure gladnes for the sight of her maiesties person, at ye beholding wherof, he tooke such comfort that with teares he expressed thesame. (Osgood 61 [E3^r])

The old man *could* be crying for sorrow, of course, but the queen pronounces (and *warrants*), or so the chronicler has recorded, that he is crying for gladness. Elizabeth's interpretation is championed as "gracious," and the moment is quickly subsumed into the crowd's cheering; but the equivocal tears continue. Although the event itself may not have been perfectly effected, *The Quenes Maiesties Passage* renders a nearly seamless portrayal of a queen being welcomed by a city turned out in its finest, covering over much of what is unlovely about the city with silken cloth, even stating that "the voide places of the pageant were filled with pretie sente[n]ces concerning the same matter" (55 [D4^r]); the account "contains" (in both the sense of including and controlling) some moments of gentle coercion on the part of the merchants, didacticism by those who presented the Bible, and nostalgic yearning on the part of the old Alderman whose tears are "warranted" by the queen to be tears of joy. Like a "view" it focuses on the positive features, reinterpreting them, "gilding" them if necessary, and moving on.

Turning from this ceremonial pageant to Isabella Whitney's later presentation, we can see that Whitney uncovers what is covered up and hidden in the coronation pageant as we move from the seen to the unseen, from the *scene* of the pageants to the *obscene* of the back streets. Her London is also busy "clothing" itself—but this time in the quotidian production of the various trades that Whitney enumerates rather than in the ceremonial decoration that closes off our view. I will focus on Whitney's "Will" as a document in the cartographic realm, on her use of the format of a will, and on the subjectivity of poet, woman, and "surveyor" to show how those three roles jar in early modern England.

Isabella Whitney appended what has come to be known as her "Wyll and Testament" to the city of London to her collection of 110 poems entitled *A Sweet Nosegay* (1573). The "plot" of the poem is partially summarized in the complete title:

> The Aucthor (though loth to leave the Citie) upon her Friendes procurement, is constrained to departe: wherefore (she fayneth as she would die) and maketh her VVyll and testament, as foloweth: VVithe large Legacies of such Goods and riches, which she moste aboundantly hath left behind her: and therof maketh LONDON sole executor to se[e] her Legacies performed. (E2^r)

Whitney tells us in the opening stanzas that she is "whole in body, and in minde, / but very *weake in Purse*" (51–52, emphasis added), an ironic statement for one about to make a will, since it is usually those "strong in purse" who have something worthwhile to bequeath and whose last wills would command the reader's interest. Nevertheless, this self-proclaimed penurious woman proceeds, in the language of a will, to bequeath parts of London to its citizens. Like the city maps discussed at the beginning of this chapter, Whitney's "Will" highlights major features of the city such as St. Paul's, the Inns of Court, Tower Hill, and many named streets, and minimizes other features, such as housing tracts, with a sort of symbolic shorthand. She specifies the remarkable and generalizes the ordinary: thus, many streets and buildings are named, but others are glossed over in phrases such as "In many places, Shops are full" (157) and "near the same, I houses leave, / for people to repayre" (171–2). Clearly, Whitney's "map" does not celebrate an ideal city, nor does it advance the symbolic meanings that city maps usually convey;[29] instead, Whitney addresses the prevalent themes of city maps in a satiric manner. While maps usually make some sort of claim to power—either ownership or sovereignty—Whitney's poetic version of a city map claims only powerlessness and poverty, highlighting especially the inequities of the distribution of wealth, oddly at the same moment that she stakes an ownership claim—that is, the power to will real estate and to redistribute that wealth. Although she is being ironic in doing so, Whitney's impulse to inventory or survey a property prior to allocating it in a will is a thoroughly legitimate one—one in which all of London comes to resemble an "estate" that Whitney owns.

To make her point, Whitney appropriates two predominantly male poetic genres: the complaint and the blazon.[30] Although the traditional complaint poem rued the mistreatment of a hapless, but sincere, lover in pursuit of an unattainable lady, Whitney, instead, personifies an entire city as the instigator of her woes. She finds London to be truly a town without "pitie" (15), as she expounds satirically on the wrongs she has suffered at the city's hands:

> And now hath time me put in mind,
>   of thy great cruelnes[s]:
> That neuer once a help wold finde,
>   to ease me in distres.
> Thou neuer yet, woldst credit geue
>   to boord me for a yeare:
> Nor with Apparell me releue
>   except thou payed weare.
> No, no thou neuer didst me good,
>   nor never wilt I know:
>
> (28–37)

Her complaints are peculiar, though, since they deal primarily with getting something for nothing: getting apparel without payment and room and board for a year on credit.

In addition to making use of the conventions of the complaint, Whitney's description partitions the city and proceeds to describe it as if it were a body in a manner similar to a blazon.[31] The traditional blazon works by making linguistic partition of a woman's body—praising a hand, a foot, a lip, an eye (to borrow from Shakespeare's Sonnet 106).[32] In Whitney's poem St. Paul's is "to the head" (77), followed by what might be seen as London's alimentary parts (in those areas providing specific foods, such as Bread Street, New Fish Street, Friday Street). Unlike the traditional blazon, however, Whitney's rendition attempts to clothe the body and satisfy its needs rather than merely describing it—a deliberate "correction" of the sexist blazon. Once the body is fed, it is clothed by those areas providing cloth and clothing (such as Watlying Street, Candlewyck Street, Birchin Lane, and Mercer's Hall), and perhaps even provided with satisfaction of sexual needs (in that "certayne hole, and little ease within" of the Counter Prison [line 197]).[33]

Beyond representing or *re*-presenting a body, then, this poem supplies bodily needs. The city of the poem feeds, clothes, educates, enlightens, and provides salvation and pleasure for the London citizenry.[34] Luxury items are provided (such as silver, gold, and jewels),[35] as are beds, swords and bucklers, robes for the Inns of Court, and other material needs.[36] Whitney bequeaths physicians for the sick (145); roysters and ruffians for the streets, along with surgeons and plasters to fix them;[37] tennis courts, dancing schools, and players to entertain.[38] She charts streets, buildings, institutions, and people. Whitney's description of London's streets, districts, and busy denizens also resembles particular early maps of the city. In the parts of the Copperplate Map that have survived we can see the fashion for animating the map with local figures engaged in quotidian and recreational activities: drying cloth in Moorfield, playing with a dog, practicing archery. Likewise, as we have seen, in Braun and Hogenberg's map of London a few citizens of the merchant class gather outside the city. Whitney seems to adhere to Georg Braun's requirement that "towns should be drawn in such a manner that the viewer can look into all the roads and streets and see also all the buildings and open spaces." As she measures out the city in her poetic lines, she moves from the center (St. Paul's) to the margins of the map: the city's gates, its prisons (the Tower, Fleet Prison, Counter, Newgate), and outskirts (places such as Smithfield).

If the body of the poem can be said to re-present London as a body, then it becomes quickly apparent that Whitney seems to tarry longest at those bodily parts that deal with the filth of civilization—the prisons holding those who have been expelled beyond the city's walls. It is not surprising that she would do so because of her own marginal position as a woman, a writer, and a kind of "surveyor," along with her personal anxiety regarding debtors' prisons tied to her own penurious state and her thwarted efforts, according to the poem, to secure any credit for herself that would have forestalled her current expulsion. If Whitney's journey represents a variation on the blazon, which fragments the body of London by highlighting particular neighborhoods through their specialized products (or the part of the body they serve), we can also say that Whitney's London is fragmented in a division of labor of the type that Marx would later conclude "reduces the labourer to a fragment of a person" (David Harvey 104). By becoming identified primarily with that fragment of the body for which their labor serves, the working people of London become metonymically mere "hands" or appendages to a giant "machine." The trade unions, or "mysteries," have indeed been *de*-mystified by Whitney as she refuses to gloss over the fetishism of the marketplace, choosing instead to characterize London throughout the poem as a place where the exchange of money is paramount and where the lack of money can only create alienation and expulsion.

By enacting a "survey" of London, Whitney casts herself into a male profession—one that would, perhaps, earn her the money she lacks, but one (among many) that is barred to women. But the profitability of surveying is no more assured than that of writing. The surveyor was frequently compensated poorly, begrudgingly, or not at all, as demonstrated in letters begging for payment and citing promises of same, such as this one of 1598 from John Norden to the queen:

To the gracious consideration of the queen's most excellent majesty:

Right gracious sovereign, I cannot but humbly exhibit these my simple endeavors unto your highness' most princely consideration.

I was drawn unto them by honorable councillors and warranted by your royal favor.

I was promised sufficient allowance and in hope thereof only I proceeded. And by attendance on the cause and by travail in the business, I have spent above a thousand marks and five years' time.

By which, being dangerously indebted, much grieved, and my family distressed, I have no other refuge but to fly unto your majesty's never failing bounty for relief.

The right honorable lord treasurer hath thrice signified his good conceit of the work and of my deservings under his hand unto your majesty. Only your majesty's princely favor is my hope, without which I myself most miserably perish, my family in penury and the work unperformed, which, being effected, shall be profitable and a glory to this your most admired empire.

I endeavor to do your majesty service. I pray for your highness unfeignedly. Quid ego miser ultra.

Your majesty's most loyal distressed subject.
J. Norden[39]

In addition to the problem of payment, the surveyor was sometimes besieged by barking dogs and howling patrons—or at least so Laurence Nowell would have us believe: his map of England portrays a barking dog in one corner and an impatient-looking patron (Lord Burghley) in the other.[40] Surveying was subject to the vagaries of taste and marketplace demands. It is a profession that, as late as 1596, seems not to have been widely accepted, judging from Norden's *Surveiors Dialogue* of that year, in which the surveyor quibbles with a landowner and a farmer about the value to each of a professional survey, as we have seen. Like women, surveyors and mapmakers often find themselves at the margins of society.[41] Given the evidence about the lot of the mapmaker and surveyor, then, we can't help wondering if Whitney would not be equally grieved if she were permitted to pursue this alternate career.

A writing career also posed problems for a woman of Whitney's day. In fact, publication carried a stigma in the sixteenth century for English men as well as women.[42] Coterie circulation was the more dignified means of transmitting one's poetry. But eventually, the stigma of print faded, and even dignified poets published their work. For women, however, the stigma of print was more pernicious and persistent; since volubility was associated with harlotry, women dared not speak in public, much less allow their words to appear and circulate in print.[43] Consider Thomas More's letter praising his brilliant daughter Margaret:

> In your modesty you do not seek for the praise of the public, nor value it overmuch even if you receive it, but because of the great love you bear us, you regard us—your husband and myself—as a sufficiently large circle of readers for all that you write.[44]

More makes it clear that women's writing, like women themselves, should be confined to hearth and home.[45] In fact, the text circulating in public streets and being exchanged for money is frequently accorded the status of the harlot. And Whitney's text, by calling attention to the streets themselves through which poet, text, and reader are now circulating physically and metaphorically, underscores the transgressiveness of this particular text and author. Whitney puts herself on the map of London and encourages (and even taunts) the reader to remember her association with its streets even after she is "gone." She also shamelessly advertises the book and the bookstalls that will sell her book within the poem.

> To all the Bookebinders by Paulles
>   because I lyke their Arte:
> They e[ve]ry weeke shal mony have,
>   when they from Bookes departe.
> Amongst them all, my Printer must,
>   haue somwhat to his share:
> I wyll my friends these Bookes to bye
>   of him, with other ware.
>
> (243–250)[46]

In Elizabethan England, a woman's reputation was on the line if she published her work, with one notable exception—the Mother's Legacy or *ars moriendi*. Here, the woman about to die is conferred a certain amount of power by virtue of her nearness to the spiritual afterlife, and is at last accorded the privilege of writing with seeming impunity. The device of the Mother's Legacy was a fashion born of necessity: the imminence of possible death in childbirth prompted a woman to write words of advice to the child she might never know and for whom she might soon lose her own life. The initial publication of a few such documents (posthumously, by a woman's surviving husband) gave rise to the eventual more widespread use of the device of the Legacy as an entrée to publication even during the woman's lifetime. Ironically, the position of power from which the woman speaks in these documents is based on complete powerlessness—complete erasure. Intended to facilitate a relationship beyond the grave, the Legacy is dependent on the testator being dead and already without property rights, except perhaps to a cemetery plot— but certainly *not* to the whole city of London. Because Whitney's poem is in the form of a will, it too is dependent on her absence and on the fiction of her death. As Wendy Wall points out, "the form of the last testament allowed women to participate in generational transmission as well as affording them an arena in which the legal/economic power denied by the culture could be simulated" in the passing on of wisdom and spiritual knowledge (1991, 45). Although spiritual wills as well as mock wills were popular at this time, and although Whitney is parodying this form and manipulating it to her own ends, she is nevertheless presumptuous in staking even a satiric claim to real property and acting as if she owns and has the right to will parts of the city to various persons. For one thing, the right of a woman to own property at all, let alone to will it without the permission of her husband, was a hotly contested issue in the late sixteenth century.[47]

Through a displaced poetic persona Whitney challenges the place of women in society as silent and obedient wives and mothers, while carving out and willing new places for women *and* for the poor within the physical space and societal limitations of sixteenth-century London. The "paper landscape" version of London to which Whitney's poem labors to give birth is a London that is kinder to the poor and the downtrodden.[48] Unfortunately, Whitney doubly exiles herself at the same time by departing London and by writing from the marginal, but oddly enabling, state of approaching death—even though in her case the approaching death is a fiction (the title tells us, after all that "she fayneth as she would die"). Whitney is bold in claiming a public voice as a woman, in publishing what is thought to be the first book of poems by an English woman, in claiming the power to own and to will property, and in entering the male professions of writer and, metaphorically at least, surveyor. But perhaps in order to appease the orthodox discriminations of the censoring arm of the Stationers' Register, this kind of audacity must be punished by the poem's conclusion. So it is that finally, in a poem that deals with social space and doles out cartographic place, Whitney ends up impoverished, self-exiled, and placeless—willing *herself* not even a tomb of her own, but instead a small, unmarked grave:

> And though I nothing named haue,
>   to bury mee withall:
> Consider that aboue the ground,
>   annoyance bee I shall.
> And let me haue a shrowding Sheete
>   to couer me from shame:
> And in obliuyon bury mee
>   and never more mee name.
>
> (311–318)

This passage is filled with irony. Here, finally, after more than three hundred lines of giving away what isn't hers to give, Whitney asks for something for herself. It seems as if she has willed away the power she has claimed throughout the poem to bequeath, and as she approaches the liminal state of her fictional death must finally ask for something in return—though only an unmarked grave so that her body will not cause "annoyance" above ground and a "shrowding Sheete" to cover her from "shame." Although her poem has opened the readers' eyes to much of the city that was covered over in the *Quenes Maiesties Passage,* Whitney finally wishes to cover her own shame as an indigent woman with a shrouding sheet. Even though she purportedly wants the city to "Rejoyce in God that I am gon" (323), she obviously doesn't want the "oblivion" she solicits here since she has advertised her book for sale earlier in the poem (243–250).[49]

In Whitney's poem we see behind the seamless façade erected for Queen Elizabeth as Whitney lays bare the failure of some of Elizabeth's promises to the poor. Although some of Whitney's Londoners seem busy (working in their shops or partaking in recreational activities), even happy (being entertained), the parts of London that are glossed over and draped off in Elizabeth's pageant are exposed here. We see in Whitney's dilatory and seemingly unplanned route through the London of her poem the utter harborlessness of one woman's city experience—for her there is no safe haven; she seems to be compelled to keep moving, however haphazardly. In contrast to the small number of stops made by the queen (nine); Whitney stops at, or wills, more than thirty-two places. And as Whitney moves circuitously through the city, it seems to chew her up and spit her out—or perhaps more accurately, as she wends her way through its system, she ends up in the bowels of the city (the prisons) with the other refuse.[50] Ironically, she cannot even claim a place for herself here, since, as she points out, a town that never gave her any "credit" cannot very well dispatch her to debtor's prison. But her plea of poverty, whether true or merely a satiric device, makes us question how London treats its poor and how it regards its workers. Whitney's London seems hostile to its own people, in stark contrast to the London that presented itself to Elizabeth not fifteen years earlier. London is hospitable to the queen, as she passes through—perhaps *because* she is only passing through.[51] But to this other woman, who arrives, as far as we know, without a plan, and more importantly without title or money, it is hostile (at least according to her poem). In the end, it might be that whether the streets of London are turned out in their best for the queen or in their worst for Whitney, there is simply no place for an unmarried woman of any rank. She can move with ease through streets covered with satin, so long as she is gone within the day, or she can roam about from place to place for longer terms, but eventually she will end up in the outskirts, beyond the walls, forced to keep moving.[52]

Steven Mullaney's impression of the city fits the *view* that Elizabeth's pageant gave us: as she moves through the city, she marks and creates a place of history. Mullaney's city, it seems, is shaped by ceremony and read primarily for its traces of that ceremony. Buildings, streets, and other places are viewed in terms of their historical associations. I would contend, however, that while these "historical" elements are important and certainly are available to be read in the topography, the diurnal activities of commoners are also important to the topology, and these activities are what the Copperplate Map, the Woodcut Map, and others clearly animate. These maps also contain a reminder of the monarchy in the royal barge on the river; however, the citizenry who are engaged in their daily rounds far outnumber and are far more prominent on these maps than is this reminder of regal ceremony (although the reminder of the royal presence is important, as we have seen). Like a "view," Mullaney's reading tends to discount the daily activities of ordinary people in defining the city, privileging instead ritual and ceremonial activities, be they civic or ecclesiastical. He remarks, for instance, that in "the secular half of the year from Midsummer's Eve to the initiation of Christmas festivities, there was little public ceremony" (20), and he goes on to discuss the other six months dominated by the Church calendar in terms of ceremony and ritual progresses that work toward "defining the terrain of community and charting, in all its particulars, the province of communal concerns" (20). Mullaney's bias is especially acute when he claims that "[p]rior to the sixteenth century, the inscriptive and interpretive process of civic ritual was sufficient to its object." Of Stow's "*Survay*" [*sic*] he comments that

> Fifty years before its appearance in 1598, Stow's *Survay of London* would have been superfluous. The city was an emblem of cultural identity and security, a symbolic text that was both inscribed by the *passage of power* and *communal spectacle*, and interpreted or made accessible through such *ritual processes.*
>
> (14, emphases added)

Again, the implication is that ceremony, the "passage of power" and "communal spectacle," alone defines the city and that every citizen will read London (the "symbolic text" above) in the same way. Mullaney usefully considers that Stow's London "serves its surveyor as a vast memory system, an extensive memory theater. His passage through it amounts to an attentive transcription of the memory-traces impressed upon the city by time and ceremonial circumstance" (16)—similar to the wax tablets discussed earlier. Mullaney is quick to differentiate this memory theater from those artificial memory theaters discussed by Frances Yates, because this "is also a city: a symbolic and even a rhetorical device, but an inhabited one" (16). But it is also useful to consider how the same city serves as a memory theater for all of its citizens, and not just for "ceremonial circumstance."

Using something like Elizabeth's progress as a defining moment for the city makes the city an apparent *tabula rasa* prior to her appearance—which is certainly not the case; instead, her ceremonial progress simply adds another layer of meaning to locales that were already loaded with meaning, if not for her, then certainly for the citizens. Conversely, Whitney's description of the common places of everyday London creates a quite different impression. Tellingly, Mullaney begins his analysis by looking at a nineteenth-century engraving based on Anthony van den Wyngaerde's Panorama of London of 1543; the city he defines, like the panorama with which he begins, is indeed the kind of portrayal we get from a "view." Whitney's poem, the early maps, and city comedy, to which I will now turn, prove Mullaney's city defined by ceremony and ritual alone to be inadequate.

While Whitney's poem purports to encapsulate the lived experience of one sixteenth-century woman, and while royal pageantry adds to the collective memory of the city and works to create a sense of place, city comedy utilizes both frames of reference and works not so much to augment the collective memory, or to improve individual memory as the artificial memory of the memory theater does, but to deploy memory in the service of satire. A certain knowledge of parts of the city, their denizens, and their associated valences is necessary to understand these plays and the comic and satiric associations evoked in the naming of specific places in the plays. Members of the contemporary audience would bring some version of an individual mental map of the city to the theater; many would have walked these streets on their way to the performance and in their quotidian rounds, others would have traversed them in carts and coaches.

John Gillies argues convincingly for a dialogic relationship between Shakespeare's Globe Theatre and Abraham Ortelius's cosmography in which "each reads the other." "All the world's a stage" was a commonplace before Jaques made his now famous speech in *As You Like It,* and Ortelius's first atlas, the *Theatrum Orbis Terrarum,* makes the relationship clear. Atlases that followed also made use of the word *Theatrum* or *Speculum* in their titles: atlases were "theaters," "mirrors," or "glasses." Ortelius himself explains the theatrical nature of the maps in his atlas:

> These Chartes being placed, as it were certaine glasses before our eyes, will the longer be kept in memory, and make the deeper impression in us: by which meanes it commeth to passe, that now we do seeme to perceive some fruit of that which we have read. . . . The reading of Histories doeth both seeme to be much more pleasant, and in deed so it is, when the Mappe being layed before our eyes, we may behold things done, or places where they were done, as if they were at this time present and in doing. (qtd. in Gillies 71)

The passage refers not just to the theatricality of the maps but also to their memorial applications; points on maps can be used as *loci* upon which to remember history just as features of memory devices and memory theaters could be used to anchor bits of information to be remembered. Many contemporary maps include famous battle scenes on their faces or in their margins and act as *aides-mémoire* for the study of geography *and* history. Gillies argues convincingly that the stage operates like a map of the world; this analogy is also applicable to city comedy where the stage operates like a city map laid out before the audience: where Smithfield and Cheapside are brought onto stages in Bankside by means of metaphor and representation. The entire city is not, of course, mapped on the stage, but the parts that are brought on "in little" are located and contextualized by the playwright, the actors, and audience members based on their mental pictures of the map of the city, as well as its human geography. I would argue that when transmuted to the local level, famous events that serve as decoration on national maps are replaced by the daily activities or typical scenes that are inscribed on city maps: a man greets a woman in the foreground of Braun and Hogenberg's map of London; a number of people meet in Smithfield and Finsbury Field while others tend to the business of drying newly dyed cloth in Moorfield in the Copperplate Map; several groups of country folk enter London over the London Bridge in Visscher's "View" of London, while others loiter around the Globe Theatre and peek into "The Bear Gardne" [*sic*]. These and many other ordinary events are also brought to life on the stages of the day, where they are spread maplike before an audience that has a bird's-eye view of events "layed before [their] eyes" that they "may behold things done, or places where they were done, as if they were at this time present and in doing," just as Ortelius would wish.

The medieval distinction between the *locus* and the *platea* of the stage is also useful to this discussion. Robert Weimann utilizes this distinction in his discussion of Shakespeare's dramas where the *locus* is used to depict specific actions that occur in specific locales, and the *platea* is used for more general commentary upon those actions—commentary that relates the specific action to more universal themes and is therefore delivered from the unlocalized part of the stage nearer the audience.[53] We might then envision some of the action of city comedy, that for which there is a specific locale (such as St. Paul's, Cheapside, Smithfield, or Quomodo's shop), taking place on the *locus* of the stage and being fixed and anchored there, and the more generalized actions (comments on the city in general and on the differences between it and the country, for example) occurring on the *platea*.[54] A useful connection can then be made between the *locus* of the stage and the *loci* of the memory theater where specific bits of information to be remembered are anchored. In city comedy, those bits of information are streets, buildings, churches, shops, and the like. Visually, Braun and Hogenberg capture this same distinction. In the foreground, or what we might consider the *platea,* the men and women meet in an unspecified setting: we don't know exactly who they are (except that they are of the merchant class) nor do we know exactly where they are (except that they are somewhere on the south side of the Thames with London behind them); that their manner of dress inscribes them as members of the merchant class provides the viewer with a comment on the city map behind them. Carrying the analogy further, we see that the actual map portion occupies the *locus,* specific places are visible, and the actions that take place at these *loci* would presumably be like those actions in the city comedies that are anchored to specific locations.[55] With the Braun and Hogenberg map, these streets might also be used as *loci* for placing the streets and landmarks in memory; the memory theater or *aide-mémoire* that is thereby produced becomes a backdrop against which audience members view the action of the drama on stage in the city comedies. The audience might be further involved (and implicated) in the action of the play if, as Weimann suggests of Shakespearean drama, some of the dramatic action took place in the yard or the pit. The "map" on stage is then extended into the audience and to the very streets to which cast members and audience will shortly return, making all the *city* a stage.[56]

City comedy, which thrived between 1597–1616, satirized city life and critiqued especially the merchant class and the materialism that this class embodied. Gail Kern Paster has attributed a predatory appetite to the city in these comedies, and this appetite resides most conspicuously in the merchant class. City comedy combines elements of medieval morality plays, Italian popular comedy, and prose narratives (such as coney-catching pamphlets). Ben Jonson aptly captures a sense of the subject matter, character types, and tone in his Prologue to *The Alchemist:*

> Our Scene is London, 'cause we would make known,
> No countries mirth is better than our own.
> No clime breeds better matter, for your whore,
> Bawd, squire, imposter, many persons more,
> Whose manners, now call'd humors, feed the stage:
> And which have still been subject for the rage
> Or spleene of comick-writers.

$$(5–11)$$

Theodore Leinwand argues that the genre of city comedy arose during this period because of new configurations of status among several groups that emerged as English commerce became increasingly centered in London.[57] City comedy puts the merchant class in the foreground just as Braun and Hogenberg do in their map. According to Leinwand, "City comedy moved onto the London stages when Londoners were beginning to reflect on their own discussions of social roles and their urban setting. It is a self-conscious genre whose historical specificity enters it into an ongoing debate" (1986, 43). London became a center of trade and finance as well as the center of government and law; in this very status-conscious society, Leinwand claims, "plays enacted, exaggerated, parodied, questioned, or endorsed what was already common coin" (4).

I have selected two plays that are typical of the genre in theme, characterization, and plotting and that not only highlight particular streets and specific locales but also depend on those locations as points of intrigue, recognition, or reversal in the plot. While the action of city comedies generally moves around and within the city itself, *Michaelmas Term* (1605) and *Bartholomew Fair* (1614) also convey the reader/viewer back and forth between the city, its outskirts, and the country and by so doing help to define the significance of each. *Michaelmas Term* draws a pointed portrait of the merchant class and its dubious business practices, deals at length with the theme of insider and outsider, and features the common plot of cuckoldry. *Bartholomew Fair* deals with issues of boundary crossing, representation, antitheatricalism, citizens' behavior, misbehavior, and the control of same.

Thomas Middleton's *Michaelmas Term* takes as its theme the attempted fleecing of a country squire of his land, the seduction of a city wife, and the concomitant cuckolding of her husband. Performed in 1607 by the "Children of Paules," the quarto edition was printed the same year without author attribution. It was first assigned to Middleton, it seems, in 1658.[58] In what will become a series of comparisons between the country and the city, the play opens with the suggestion that coming to the city necessitates that one shed the distinctive clothing of the country. Part of that distinctive "clothing" of the country, it seems, is one's estate in land, but another part is one's conscience, as the personified "Michaelmas Term" makes clear in his first speech of the *Inductio:*

> Lay by my conscience,
> Give me my gown, that weed is for the country;
> We must be civil now, and match our evil;
> Who first made civil black, he pleas'd the devil.

. . . .
And so through wealthy variance and fat brawl,
The barn is made but steward to the hall.

(Ind. 1–12)

In *Michaelmas Term* at least, an essential part of the clothing or "weed" of the country is the conscience; this weed has no use in a city where "civil" is coupled with "evil" and "devil" (doubly emphasized in the internal and end rhymes of lines 3 and 4) and with the black of the robe itself (4), the color of the devil. Because of the "civil evil" of a city without a conscience the country gentleman in this play, Richard Easy, is easily gulled into pledging his land for the trinkets and baubles of the city, such as a bolt of cloth. This sartorial theme is maintained throughout the play as characters dress for the various roles they play in the central plot.[59]

Ephestian Quomodo, a city woolen draper, makes it clear that for him land—an estate in the country, and especially the income it produces—is more important than any woman, including his own wife, or her marital fidelity. In fact, for Quomodo, having the one facilitates obtaining the other:

> There are means and ways enow to hook in gentry,
> Besides our deadly enmity, which thus stands:
> They're busy 'bout our wives, we 'bout their lands.

(1.1.105–107)

Each wants what the other has, it seems: as the country gentlemen are "hook[ed] in" by the wives of the merchants, the merchants will be "busy" duping them out of their lands. Shortyard, one of Quomodo's accomplices, seconds this sentiment by adding that "To be a cuckold is but for one life, / When land remains to you, your heir, or wife" (1.1.109–110). The syntax here suggests that the land not only compensates for the loss of one's wife in the previous passage, but actually *is* heir and wife if we read "your heir, or wife" as an appositive. I suggested a reading of "To Penshurst" that privileges the income potential of the various subdivisions of the estate as they are named and described, but that poem never approaches the bluntness of Quomodo's avaricious prosaic pronouncements, "Now I begin to set one foot upon the land. Methinks I am felling of trees already; we shall have some Essex logs yet to keep Christmas with, and that's a comfort" (2.3.338–340).[60] A similar sentiment underlies Quomodo's projecting the envy of others at the "goodly load of logs" and the "pleasant fruit" from the orchard he imagines he possesses (3.4.15–17); a "country house poem" of Quomodo's creation would overturn any of the subtlety we have just examined in Jonson's poem. And even after he himself has been outdone by Richard Easy, Quomodo finds comfort enough in the land to compensate for his newly acquired cuckold's horns:

> He does devise all means to make me mad,
> That I may no more lie with my wife
> In perfect memory; I know't, but yet
> The lands will maintain me in my wits;
> The land will do so much for me.
>
> (5.3.64–68)

Of course, Quomodo doesn't yet know that he has also been tricked out of the very land he tricked Easy out of earlier in the play and thus will *not* have the lands to "maintain [his] wits." Unlike the country house poem, where wealth and generation seem to be an act of providence (at one level at least), the plots in this play oppose money to fertility in such a way that possession of wealth seems to preclude procreation. As Quomodo's man Shortyard explains:

> We have neither posterity in town, nor hope for any abroad . . . We could not stand about it, sir; to get riches and children too, 'tis more than one man can do. And I am of those citizens' minds that say, let our wives make shift for children and they will, they get none of us; and I cannot think but he that has both much wealth and many children, has had more helps coming in than himself. (4.1.33–38)

In stark contrast to the country house poem, in the London of *Michaelmas Term* illegitimate children are actually hoped for abroad, and it seems that wives must actively shift for themselves in the getting of children. Shortyard, who cannot "stand about it," is frankly suspicious of anyone who has both progeny and wealth—they must have "more helps coming in than himself."[61]

Michaelmas Term presents a London in which its shopkeepers practice trickery and duplicity, and where the darkness of the shops obscures the cheating and conniving that occurs there (very unlike the merchants in Braun and Hogenberg's sunny depiction, who meet in the open). Quomodo is in fact proud that his shop is "not altogether so dark as some of [his] neighbors,' where a man may be made cuckold at one end, while he's measuring with his yard at tother" (2.3.32–35);[62] but his pride, I think, stems from his belief that he can cheat his customers just as well as his competitors can, even in this brighter venue. Ironically, in *Michaelmas Term* the darkness of the shop that is usually conducive to cheating the customer, along with other unscrupulous business practices employed by Quomodo, becomes his own undoing. In the end he is cuckolded in his own shop (while he measures with his yard) despite his shop's relative brightness.

Not surprisingly, city comedy characterizes London as having all the vices for which the theaters had been attacked in antitheatrical tracts. The practice of boys playing women and thus inscribing a kind of homoeroticism on the stage,[63] as well as the kind of shape-shifting that acting and costumes allow, were two of many theatrical practices that the Puritans found objectionable. The kind of shape-shifting that Quomodo enacts on stage in becoming a "friar" was already a common dramatic element that made the dangers of clothes switching and class switching seem more threatening to Puritans. But *Michaelmas Term* goes beyond the mere suggestion of homoeroticism implied by boys playing women with its numerous homosexual innuendoes and insinuations of sodomy along the lines of Cockstone's greeting to Easy:

> You seldom visit London, Master Easy,
> But now your father's dead, 'tis your only course;
> Here's gallants of all sizes, of all lasts;
> Here you may fit your foot, make choice of those
> Whom your affection may rejoice in.
>
> (1.1.42–46)

With all this talk of "gallants of all sizes," "lasts," "fit," and again the common pun on "foot" (for the French *foutre*), especially in the mouth of a character named "Cockstone," this passage seems to make manifest one of the Puritans' worst fears about the theater, that viewing a suggestion of sodomy would excite such desires in the viewer. Here the actors aren't even dressed as man and woman, but as man and man in the all-male domain of the Inns of Court.[64]

But this city on display is also a city that sees the need of its own purgation. As Shortyard responds crudely to toasts his friends have made:

> This Rhenish wine is like the scouring-stick to a gun, it makes the barrel clear; it has an excellent virtue, it keeps all the sinks in man and woman's body sweet in June and July; and, to say truth, if ditches were not cast once a year, and drabs once a month, there would be no abiding i'th' city. (3.1.195–200)

Presumably, this is the London with which theatergoers are most familiar (unlike the satin veneer of Elizabeth's progress), else the comedy, though frequently as unrefined as the passage above, would not work. City dwellers (of whom a representative sample is in the audience) are shown on the stage against a backdrop that a familiar "view" of the city (either physical or representational) would provide, or for which a mental map would supply a framework and necessary details. These backdrops would include the well-known places depicted on views, the equally familiar parts of the city that are not displayed in such representations, and the commonplace people and types traversing the streets, fields, and waterways of the city maps. It is not simply the topical references to streets, buildings, and landmarks that bring these plays into the realm of cartographic inquiry, but the ideologies of place that they contain—and places, like maps, are never neutral, but are always filled with meanings, culturally shared meanings as well as meanings based on personal experiences.

Ben Jonson's *Bartholomew Fair* (1614) takes us beyond the city gates to the northern outskirts of seventeenth-century London, to a fair.[65] At the center of the Fair and of Jonson's play is the booth where the pig woman, Ursula, marks out her territory by her leaky pissing and profuse sweating.[66] Indeed, a shared craving for the pork served by Ursula is the peculiar impetus for the citizens' sojourn to the fair in the first place. Win Littlewit's craving for pork (or, more accurately, her husband John's craving consigned conveniently to her womb because of her pregnancy) and the collateral fear of miscarrying if she doesn't give the baby what it wants (1.6, but really what John wants) justify the Littlewits' attendance. In contrast to the mythical "Lubberland" mentioned in the play, where pigs go around roasted and begging to be eaten, and certainly in contrast to the country house poem, here one must go out of one's way to look for food, and in the course of this quest, it seems, one is exposed to all sorts of other temptations (3.2.68). For some in the play, such as Zeal-of-the-land Busy and Dame Purecraft, boundaries are based on religious prohibition—and entrance to the world of the Fair, if not scrupulously avoided, must be carefully rationalized. Busy, a Puritan from Banbury (a place of notorious Puritanism), is quite single-minded in his insistence on remembering the purpose of their visit to the Fair. Since that purpose is to find pork, the readiest way, according to Busy, is to follow their sense of smell—to "enter the tents of the unclean for once, and satisfy your wife's frailty. Let your frail wife be satisfied: your zealous mother, and my suffering self, will also be satisfied" (3.2.75–77). Here it seems the Puritan is able to achieve what is characterized as scandalous satisfaction without admitting that he too relishes that satisfaction in the taste of the pork. Later on, he even avows that his public consumption of pork is a deliberate act intended to dispel any doubts about Puritans' association with Judaism.

The fair as microcosm of the world or of the city might better be termed a "heterocosm." The market here is of clearly tainted goods, as described by Leatherhead when he accuses Joan Trash of using "stale bread, rotten eggs, musty ginger, and dead honey" in making her gingerbread (2.2.8–9); her name, too, seems hardly an endorsement of quality. Ursula cheats her beer customers and unabashedly admits to Leatherneck that she will charge more for a pregnant woman who craves pork—"sixpence more" (2.2.100). These are the very offenses that Justice Overdo describes as "the very womb and bed of enormity" (2.2.95) and that he hopes to weed out by going around the Fair in disguise (2.1). Codes of sexual behavior, however, are different in this little world outside the city than those presumed in the city, and merely showing up here seems to have its consequences in casting aspersions on a lady's reputation, as Quarlous, Winwife's man, advises his master: "Now were a fine time for thee, Winwife, to lay aboard thy widow, thou'lt never be master of a better season, or place; she that will venture herself into the Fair, and a pig-box, will admit any assault, be assured of that" (3.2.116–9). Certain codes of behavior—certain morals—are associated with certain places on the city map, and merely appearing at the Fair, and more particularly at Ursula's booth, seems to be enough to code the widow as approachable and attemptable, if not downright lascivious and loose—one who would "admit any assault."[67]

Like the theaters, which are also in the suburbs, Bartholomew Fair is a place of transformation—therein (for many) lies the danger. It is a place where justices become madmen, where gentlewomen become "birds of the game," and where husbands may even be convinced that this type of metamorphosis is a good thing (4.5). It is a place where stoics are put in the stocks, where "they have set the faithful . . . to be wondered at; and provided *holes for the holy* of the land," according to Busy when he is impounded in them (emphasis added 4.6.110–111). It is, to quote Justice Overdo once more, "the very womb and bed of enormity": Jerusalem, the sacred center of medieval *mappaemundi,* has been replaced at the metaphorical center of this microcosm by the this "womb" and by the holes of the stocks—or perhaps more accurately, by an exchange of "holes for the holy" center of the medieval map.[68] The craving for pork that brought the pregnant Win and her entourage to the fair in the first place is quickly sated, forgotten, and replaced with a new craving for other "sights" of the Fair. But it is Win's need to find a place to urinate that brings on her temporary, but potentially dangerous, transformation into a high-priced prostitute. Smithfield is a place where even honest wives might fall prey to the convoluted logic of the likes of Jordan Knockem, who in an unflattering analogy between dunghills and women, describes the abundance of coaches at the women's beck and call should they enter a life of sexual promiscuity:

> Oh, they [coaches] are as common as wheelbarrows, where there are great dunghills. Every pettifogger's wife has 'em, for first he buys a coach, that he may marry, and then he marries that he may be made cuckold in't: for if their wives ride not to their cuckolding, they do 'em no credit. Hide, and be hidden; ride and be ridden, says the vapour of experience. (4.5.89–94)[69]

Bartholomew Fair is also a place where old entertainments like the story of Hero and Leander are reduced "to a more familiar strain for our people" (5.3.97). The action is translated to the banks of the Thames (both literally in the playhouse where *Bartholomew Fair* would have been played and dramatically in the substitution of the Thames for the Hellespont), the local Puddle Wharf is substituted for Abydos, and the Bankside for Sestos; Leander has become a dyer's son; Cupid is a drawer; and Hero is "a wench o' the Bankside, who going over one morning, to old Fish Street; Leander spies her land[ing] at Trig-stairs, and falls in love with her" (5.3.101–106). The story of Hero and Leander, in other words, becomes a city comedy. The ridiculous puppet play allows a metadramatic confrontation between one Puritan and the valiant (if ridiculous) arbiters of representation on the issue of dramatic transvestism.[70] As one of the puppets argues:

> It is your old stale argument against the players, but it will not hold against the puppets; for we have neither male nor female amongst us. And that thou may'st see, if thou wilt, like a malicious purblind zeal as thou art! (5.5.88–91)

Laura Levine reads *Bartholomew Fair,* and especially this puppet play within the play, as Jonson's direct response to the antitheatrical tracts. Levine poses the question: "Why is it that Jonson has cut from his *Hero and Leander* everything potentially objectionable to an antitheatricalist?" (91–92). As Levine describes it:

At the heart of the anti-theatrical "psychology" of homosexuality, then, is a profound contradiction about what is provoking the sexual response in the first place—man or woman. And it is as if to mediate this conflict that anti-theatricalists organized the experience of desire itself chronologically: the spectator begins by getting excited at the theatre, and ends by having sex with the man. (97)

For Levine, the puppet play then becomes a "vision of what a world would be like in which theatre had already accommodated itself to anti-theatricality" (98). If, as Stephen Gosson suggests, "garments are set down for signes distinctive betwene sexe and sexe" (175), then Jonson has severed "the relationship between sign and thing because there is no 'thing' under the sign, no genital under the costume for the sign to refer to" (Levine 101). Despite these puppets' blatant sexlessness, I am reminded of the "puppets" that Hamlet wishes to direct (*Hamlet* 3.2.246–7), and I believe this association is deliberate on Jonson's part.

Ironically, in a play where a number of bad (and superficially oxymoronic) puns are bandied about regarding the "foul Fair" and the shortage of "fowl" (or prostitutes) at the fair, Bartholomew Fair ends up being a place where the foulness of the Puritans is exposed. Dame Purecraft, it turns out, is more *crafty* than *pure* in her shady and quite lucrative matchmaking; and Zeal-of-the-land Busy is exposed as having been quite "busy" bilking fortunes from deceased brethren of his fold. The implication seems to be that vices of other "fair," "pure," or "busy"- body Puritans might also be uncovered in such a place.

Jean Howard has called the antitheatrical tracts a "genre of anxiety" (1994, 23); that anxiety is brought both to the stage and to the puppet stage in *Bartholomew Fair,* where the objections of the Puritans are overthrown because the Puritans themselves are proved to be hypocrites. One of the problems that Howard gleans from the antitheatrical tracts of John Northbrooke is that for him it seems that "place" determines identity: "people at the theater are not *where* they should be (i.e., in their parishes, at work or at worship); consequently, they are not *who* they should be, but are released into a realm of Protean shapeshifting with enormous destabilizing consequences for the social order" (27). Howard goes on to say that "Northbrooke and many subsequent antitheatricalists, along with those devising new statutes to regulate vagrancy, are obsessed with *re*marking and *re*situating the bodies of 'the idle' and 'the unmastered'" (27). In drama, of course, much of that shapeshifting occurs on stage as actors assume various roles, and more blatantly when they assume disguises within the action of a play.

City comedy not only places all of the types against which North-brooke, Phillip Stubbes, and other Puritans rail (adulterers, whore-mongers, dicers, userers, and the like) up on a stage, but mimetically locates them in specific London locales brought onto the stage. Of course, one of Stubbes's chief grievances against the theater is that people will leave and imitate these "doble dealing ambodexters" out-side the realm of the theater. He fears the lessons that might be gleaned from such entertainments:

> Of Commedies, the matter and ground is love, bawdrie, cosenage, flattery, whordome, adulterie: the persons or agents, whores, queanes, bawdes, scullions, knaues, Curtezans, lecherous old men, amorous yong men, with such like of infinit varietie: I say if there were nothing els, but this, it were sufficient to withdraw a good christian from the using of them. (L7^{r-v})

Here it almost seems as if Stubbes is ticking off the characters of the city comedies we have just examined. Stubbes continues his warning by turning more specifically to the kinds of behavior that are por-trayed in the theater—behavior that he believes will likely be imitated upon leaving a play:

> Do they not maintaine bawdrie, insinuat folery, & renue ye remem-brance of hethen ydolatrie? Do they not induce whordome & un-cleannes? nay, are they not rather plaine deuourers of maydenly virginitie and chastitie? . . . Than these goodly pageants being done, euery mate sorts to his mate, euery one bringes another homeward of their way verye friendly, and in their secret conclaues (couertly) they play ye Sodomits, or worse. And these be the fruits of Playes and En-terluds, for the most part. (L8^{r-v})

Although Stubbes wrote this treatise years before the plays I have just discussed, his treatise could practically be an advertisement for the ac-tion of these plays. The great fear stated above is that the actions de-picted on the stage as occurring on the streets and in the households of London will be reenacted on those very streets by those attending the theaters—in fact, by "every mate" and "every one" who at-tended. The theater couldn't portray such vices if they didn't exist to some extent in the city already, but the fear expressed in the antithe-atrical tracts seems to be not so much that a play like *Michaelmas Term* represents an accurate map, but that it will produce the kind of

behavior in the city that will make it *become* an accurate map. And
that playgoers will learn the lessons from this "school" that Stubbes
enumerates at length:

> if you will learne falshood, if you will learn cosenage: if you will learn
> to deceive; if you will learn to play the Hipocrit: to cogge, lye and fal-
> sifie: if you will learn to iest, laugh and fleer, to grin, to nodd, and
> mow: if you will learn to playe the vice, to swear, teare, and blaspheme,
> both Heaven and Earth: If you will learn to become a bawde, un-
> cleane, and to deverginat Mayds, to deflour honest Wyves: if you will
> learn to murther, slate, kill, picke, steal, robbe and rove: If you will
> learn to rebel against Princes, to comit treasons, to consume treasurs,
> to practise ydlenes, to sing and talke of bawdie loue and venery: if you
> will lerne to deride, scoffe, mock & flowt, to flatter & smooth: If you
> will learn to play the whore-maister, the glutton, Drunkard, or inces-
> tuous person: if you will learn to become proude, hawtie & arrogant:
> and finally, if you will learne to comtemne GOD and al his lawes, to
> care neither for heauen nor hel, and to commit al kinde of sinne and
> mischeef you need to goe to no other schoole, for all these good Ex-
> amples, may you see painted before your eyes in enterludes and playes.
> (L8ᵛ-M1ʳ)

In the opening pages of his *Anatomy of Abuses,* Stubbes sets himself up as a traveler and describes why he enjoys travel:

> Truely, to see fashions, to acquainte my selfe with the natures, quali-
> ties, properties, and conditions of all men, to breake my selfe to the
> worlde, to learne nurture, good demeanour, & cyuill behauiour: to see
> the goodly situation of Citties, Townes and Countryes, with their
> prospects, and commodities: and finally, to learne the state of all
> thinges in general: all which I could neuer haue learned in one place.
> For who so sitteth at home, euer commorante or abiding in one place,
> knoweth nothinge, in respecte of him, that trauayleth abroade: and
> hee that knoweth nothing, is lyke a brute Beaste, but hee that knoweth
> all thinges (which thinge none doeth but God alone) hee is a God
> amongest men. (B1ᵛ-B2ʳ)

Although Stubbes would not admit it, his rationale for traveling
could also be used to justify attending the theater. And certainly his
remarks are similar to Sir Thomas Elyot's remarks regarding the use-
fulness of maps in *The Boke Named the Governor*—that one can "behold those realms, cities, seas, rivers, and mountains . . . without peril
of the sea or danger of long and painful journeys" (35). But whereas
Elyot is endorsing viewing the representations of a place as an avenue

to understanding, or even a substitute for experience, Stubbes will have none of it. Stubbes may indeed believe himself to be a traveler, but his need to see everything for himself reveals a fundamental mis-understanding of the representational devices such as the stage and the map.

In contrast to city comedies, condemned by Stubbs, some depic-tions of the city sometimes cast it and its denizens in a positive light, as we have seen, dressing it up and making it presentable in pageants for the outsider's review. Records for the city of Wells indicate that for a civic pageant presented for Queen Anne in 1613 plans were made "to decorate the streets, and rid the town 'of beggars and Rogues.'" (Bergeron 1971, 100[73]). Similar work might have been done in London for its pageants. But, even while pageants work to establish London's place in the world, and cover over that part of London that is exposed in all of its excesses in city comedies,[71] these pageants are not without hints at that seamier side of London that the city comedies exposed and exploited. In the Lord Mayor's Show of 1613, entitled *The Triumphs of Truth,* an allegorical and personi-fied Error tries to get a toehold (and, in fact, has a very long speech that could potentially dominate or steal the show)[72]; Gluttony, Sloth, and Envy make momentary appearances; and a personified "Lon-don" chides her "thankless sons"; but Zeal and Truth, of course, win out in the end (Bullen, Vol 7, 225–262).

The city provides diverse experiences and is the subject of a range of representations. In pictorial or theatric representations of the city, decisions must be made about what will be depicted and what will be left out. For the Londoner of the late sixteenth and early seventeenth century, those representations might include both the ceremonial and the diurnal: the queen's coronation progress as well as the dry-ing of cloth in Moorfield. They might depict the appetite for sex and money that shifting class structures facilitated, the common themes of cuckolding of husbands and of deceptive business practices that cheat outsiders of their possessions. Those representations sometimes depicted London as a "town without pity" that expels those who would be its citizens, sometimes as a place that had seen better days. Representations of the city might be tawdry or gilded, depending on the perspective of the mapmaker, the artist, or the writer of the poem, play, or survey. But most depictions of early modern London will, like Braun and Hogenberg's map, put members of the merchant class in the foreground, dominating the landscape of the city.

Epilogue

My subject here has been maps and memory. Cartographic place has been shaped in many ways to create "a sense of place" in the works I have discussed. I have shown myriad depictions of place that reflect a multiplicity of relationships to a variety of locales. I have moved in my discussion from the international and national levels to the local level in order to show how this same pattern of national-to-local played out in the cartographic projects and the literature of the early modern period in England. As world maps were revised to include the "New World," England had to vie for position in a now expanded world, and in what was to become a very "new" world indeed.

Some of the great mapping projects in England during Queen Elizabeth's reign were completed at the queen's behest or at the behest (and usually at the expense) of those close to her. Her image appears conspicuously on some maps to reinforce her dominion over these areas, but also to flatter her. Looking from the vantage point of four hundred years removed, we have, of course, a very different perspective. The Ditchley Portrait of Elizabeth has been read and reread; I have offered yet another reading of it here. It was originally painted to commemorate Elizabeth's visit to Ditchley Park, which is where her feet are positioned on the map. Scholars have talked about whether the map in the portrait is of any use at all with the queen standing on it as she does, but as a way of remembering queen and country as one, it works. As a memory image it imprints itself vividly in the mind. I think Mary Carruthers is correct in stating that memory can be enhanced by recalling the details of where one first saw something or first learned something—"making each memory as much as possible into a personal occasion by imprinting emotional associations like desire and fear, pleasure or discomfort" (60). The Ditchley Portrait in the National Portrait Gallery in London is huge, stunning, unforgettable.[1] When approached from one direction it is visible from several galleries away; when approached from another direction, it can catch you off guard completely. I have tried to suggest in my reading of the Ditchley Portrait some of the ambivalent messages of both power and vulnerability that might be contained in this image, but the image itself is a powerful one. It is a compelling remembrance of this queen and her island country.

I have traced the history of some of the mapping projects of early modern England, and I have been interested in how these mapping projects are refracted in the literature. It is not surprising that Spenser, who had seen maps displayed in studies and map rooms, would turn to a pageant of maps to try to communicate a colonialist agenda using a cartographic representational register even while shrouding that communique in an allegorical register. "The Poet," he tells us "thrustesth into the middest euen where it most concerneth him, and there recoursing to the thinges forepaste, and diuining of thinges to come, maketh a pleasing Analysis of all" ("A Letter of the Authors," Hamilton, *FQ* p. 738); this colonialist agenda seems to be what "most concerneth him" in this episode. Spenser's messages are "clowdily enwrapped in Allegoricall deuises" that have allowed us a glimpse into the use of maps at court not only by those in power, but also by someone who is not in power, addressing the person who is in power. J. B. Harley has suggested that "the ideological arrows [of maps] have tended to fly largely in one direction, from the powerful to the weaker in society" ("Maps," 301), but Spenser doesn't let that stop him in what is, I think, stunning audacity. His rivers serve as fluid memory images, and one could almost imagine a school geography lesson entailing memorizing their names along with Spenser's descriptions of them: the "noble" Thames, the "chaulky" Kenet, the "wanton" Lee, the "stately" Severne, and the "fatal" Welland. This episode that is saturated with rape also provides the queen with a memorable lesson about the vulnerability of her body and her country: it hints at concerns over succession that were weighing heavy at this time, and finally it provides a lesson in how the queen and her country will be remembered.

Shakespeare's Jachimo, as outsider to England, gives us an ingenious improvisation using maps and memory. His account demonstrates a failure of maps to convince, or at least a failure of maps by themselves, and a triumph of memory and tales. But the discussion in this chapter has also revealed some gender issues that trouble maps and portrayals of space that were begun in the previous chapter. Perhaps because it is the subject of study, space is frequently figured as female. The space of the New World is often personified as pure, untainted, virginal, and enticing—the virgin (Virginia) ready for defloration and/or marriage. Flip-flopping this analogy, Donne and Davies map out women's bodies and use these maps to signal sexual congress and/or conquest.[2] The space of England and its sometimes metonymic Queen may be seen as coextensive and impregnable, having fortuitously fended off incursions from the Spanish Armada and foreign suitors. Imogen, another British princess who has not been so fortunate, has had her person and her chamber surveyed like a map, memorized, and published—and she has been stained in the process.

Ben Jonson's country house poem celebrating the estate of Penshurst has elements of both a modern survey and of a Rogationtide celebration, with a "wound for every landmark." The discussion here has allowed us to look at tracts about surveying in conjunction with contemporary poetry to see how the two genres work to celebrate place and to remember it. The poem is a memorial to the place and to its owner. The countryside of the estate survey and of the country house poem represents the underpinning of legitimate inheritance, which is dependent on clear title and boundaries, and on the Lady's marital chastity. The rent is represented by the fruits of the land; the children represent the fruits of the Lord. Four images linger prominently in the memory: the wall, the child reaching for the fruit, the farmers' daughters, and the poet at the center of it all. Again here, centers and peripheries are essential to our understanding: all four images remind us of delineations of boundaries, periphery, center, and bias. Jonson has his own reasons for putting himself in the center of this little map, for remembering or re-membering the scenes at Penshurst.

The space of London is often portrayed as wanton and incontinent—the prostitute or unruly wife who needs to be controlled—the swallowing womb. Whatever subtlety the country poem contained is overturned here. I have shown how the city is remembered in maps, views, and several genres of literature. Women seem to fare poorly here unless they are regal, and perhaps then only if their visit is transitory. Isabella Whitney, or at least her poetic persona, has an entirely different experience—moving from place to place, but without the sort of purpose that Elizabeth's planned progress had. She begins with St. Paul's at the head—often the center of early London Maps—then moves here and there criss-crossing the body of the city.[3] Like Whitney's poem, maps of London were populated with local denizens busy at work and recreation. I have shown how the literary portrayals in the city comedies, as well as maps like Braun and Hogenberg's put the merchant in the foreground with the city behind them as their backdrop or their stage. As in the theater, the foreground of Braun and Hogenberg's maps resembles the *platea* of the stage; the areas behind the four large figures—the map of the specific streets, resembles the *locus*. The streets are also loci in a memory system that combines maps and memory. Both the stage and the maps serve to etch these images on the memory.

During the early modern period maps were becoming more widely available and were beginning to be used as decorative items—on tapestries, in paintings, and even on playing cards. One deck of playing cards that has survived from 1595, has one of the fifty-two counties of England and Wales on each of the fifty-two cards. These maps are accompanied by information about each of the counties, including length, breadth, and circuit; information about terrain, chief products, such as, "HVNTINGTON fitt for corne and cattell / Fenny & plentifull with plesaunt hills and groves"; designations of neighbor counties or bodies of water are also included. The maps are all of the same scale and orientation, and because the counties are shown independent of each other the maps look almost like little *isolario* maps. These cards might be used for gaming or for memory. Besides the counties on the numbered cards, there is a card commemorating the queen in both verse and portrait, and one celebrating her ancestors; there is a card with a map of London, a card praising the city in verse, and a composite map of England with all its counties reassembled. These are accompanied by doggerel verse, above and below the images, but their use as either memory cards or as gaming cards is of interest because they represent yet another way of associating maps and memory. In an essay celebrating the influence of Christopher Saxton and the "lasting image" of his work through these cards, among other things, Victor Morgan mentions the "delight in maps as objects for aesthetic and intellectual contemplation" (404). County maps on playing cards would certainly be a delightful object to hold, study, and manipulate in any number of different ways. Perhaps because of this delight, county map playing cards were produced throughout the seventeenth century.[4] The appeal of maps in general is summed up well by Lawrence Manley in his meditation on Thomas Elyot: "This pleasurable sense of power—of appropriating a vast public space to the private recess of the study and the mind—was greatly enhanced by the typographical revolution and its treatment of the printed book as a container of knowledge readily mastered and possessed" (*Literature and Culture*, 135). Pleasure, and a pleasurable sense of power, would certainly be applicable to viewing and manipulating these playing cards.

In his discussion of the influence of Christopher Saxton, Victor Morgan was among the earliest to suggest more be done "to examine people's subjective organization of space in the past" (406). I have done so here. I have tried to show that maps, though frequently viewed as unbiased documents, are never neutral, and neither are the linguistic representations of place that I have examined alongside them. The increasing emphasis on accuracy in the technical aspects of surveying and mapmaking processes has a correlative pattern in the concern for correctness and correction in written works dealing with maps, alluding to them, or containing metaphorical allusions to the process of mapmaking; the purity of the New World might be "corrected" by conquest and rape; the wantonness of the city might be corrected by amendment of the social structure. Mapmakers and others who depict place must always choose what to include and what to exclude, what to make central and what to make marginal. In both visual and linguistic representations of the places of England the increasing concern for scientific accuracy heralds the larger problem of determining how to control the subject of representation and how to create a sense of place.

Notes

1. An alternate translation of the Borges passage is Norman Thomas di Giovanni's translation under the title "Of Exactitude in Science," *A Universal History of Infamy*, p. 141. Giovanni's translation is quite different from the one quoted above; for example, "estos Mapas Desmesurados," translated here as "those Unconscionable Maps," becomes "these Extensive maps" and "Menos Adictas" is translated as "less attentive" rather than as "less addicted." The Spanish version, "Del rigor en la ciencia," is in Jorge Luis Borges, *Historia universal de la infamia* (Buenos Aires: Emeceé Editores, 1954), p. 136.

2. The Farmer's objections center on a fear of what the map or survey might generate—higher rents, so perhaps it would be more accurate to say that he *does* see the value to maps (especially to the lord), but he sees only negative repercussions for himself and his interests.

3. David Wade Chambers makes this specific statement about David Turnbull's work in his preface to Turnbull's 1989 book (v).

   Literary criticism is but one example of the deployment of mapping paradigms to theoretical fields. A recent publication of the Modern Language Association, for example, discusses "Redrawing the Boundaries" of literary studies. It is common to use the terms "survey," "field," "margins," and "boundaries" to try to define a particular discipline or approach. See Greenblatt and Gunn as one example.

4. Harley and Woodward postulate as well that "in a sense the subject has become a prisoner of its own etymology": in some European languages (e.g., English, Polish, Spanish, and Portuguese) the word derives from the Late Latin word *mappa,* meaning a cloth; in other European languages, the word derives from the Late Latin *carta,* which means any sort of formal document (e.g., French *carte,* Italian *carta,* Russian *karta*) (I: xvi).

5. See Robert Lloyd, "A Look at Images," for an overview of the literature and a summary of the empirical research from cognitive psychology dealing with how spatial information is coded, stored, retrieved, and manipulated in the memory.

6. The word "orientation" is derived from this directional privileging.

7. Another image of Isidore of Seville's T-O map, along with several other T-O maps is available online at: <http://www.henry-davis.com/MAPS/EMwebpages/205.html>.

8. It should be noted, though, that the "O" in many T-O maps (and in the *mappaemundi* that evolve out of them) is not circular. The shapes of the maps vary from round, to oval, to square with rounded corners, as in the Beatus Map of 1109 (See Whitfield 16–17).

9. These names, in alternate spelling, appear on the T-O map shown in Figure 1.1.

10. The Hereford *Mappa Mundi* can also be viewed online at: <http://www.herefordwebpages.co.uk/mapmundi.shtml>; several images, including details and a re-drawing are found at <http://www.henry-davis.com/MAPS/EMwebpages/226.html>; a larger black and white version is found at <http://www2.vmi.edu/gen_ed/map.jpg>.

11. Again, as with the T-O maps, not all medieval *mappaemundi* are round. The "Cotton World Map" of the eleventh century is square (See P. D. A. Harvey 1996, 28–29).

12. A full color version of the Psalter Map can be viewed at the British Library website: <http://www.bl.uk/exhibitions/maps/psalter.html>.

13. The Ebstorf Map was destroyed during World War II, but photographs taken earlier have survived. The reconstruction can be viewed on the Internet at: <http://www.henry-davis.com/MAPS/EMwebpages/224c.html>; this site has several images of the reconstructed map, including close-ups and a monograph. It can also be seen at: <http://sun1.cip.fak14.uni-muenchen.de/~wallicz/ebst01.htm>.

14. Conley shows how philosophers, poets, and psychoanalysts have "idealized" the navel in terms of "ruptured attachment to the world":

> The unknown declares itself to be what escapes recognition. For the growing subject, it assures a place where sexual difference is erased. In a lexicon that evokes the process of centering and of depicting contour and relief in printed maps, the navel, both a valley and an inverted molehill (what geographers call a *taupinière*), seems to allow for the designation of a place where bodily demarcations can be effaced. It could also be a dot marking an agglomeration, a capital city, or a town. What Rosolato calls "umbilication" draws attention to a hollow (or relief), to a centering without any outlet or issue beyond, "to a 'one-eyed hole,' a hole that is not a hole, a caecum, a blind gut that represents a limit, an anti-abyss, an interruption along a path, a passage into the void through a conceptual impasse" (1978: 257). In this sense, the navel is freighted with cartographic and literary meanings (Conley 9,

the passage from Rosolato is from *La relation d'inconnu,* Paris: Gallimard).

15. Leonardo's drawing is available on the Internet at: <http://banzai.msi.umn.edu/leonardo/vinci/vitruvian.jpg> and at <http://www.leonardo2002.de/ehome/ehome.html>.

16. Edgerton suggests that this relocation implies that "the male phallus has the same symbolic significance as representing the procreative link between earth and heaven" (12).

17. For a discussion of this phrase and the Latin equivalent, "*Hic sunt dracones,*" see Erin Blake's website, "Where Be '*Here be Dragons*'?: Ubi sunt 'Hic sunt dracones'?" <http://www.maphist.nl/extra/herebedragons.html>; the phrase was actually used (in the Latin form) on only one globe—The Lenox Globe (ca. 1503–07)—although many map margins are adorned with dragons and other monstrous creatures.

18. The animals and beggars that inhabit the "mangled ruins" of the map of the empire in the Borges example of the epigraph at the beginning of this chapter is what seems to be a variation of this theme.

19. See Daniel Wallingford's map, "A New Yorker's Idea of the United States of America," shown as Figure 1.8 in Gould and White, *Mental Maps.* Wallingford's "A Bostonians' Idea of the United States," with even fewer details beyond its focus in New England (except for a few rivers and mountains), is shown as Figure 1.9 in Gould and White. Similarly, a map of "How Londoners see the North" (Gould and White, Figure 1.10) reveals the spatial biases from the viewpoint of the Doncaster and District Development Council: railways end in Manchester, only one road (the Great North Road) is shown leaving London, and the complete "end of roads" is indicated before arriving in Scotland, beyond which there are apparently only ox carts, according to this depiction.

20. Julian Yates's paper, "The Geometry of Forgetting: Maps, Mapping and the Culture of Print in Early Modern England," makes clever connections between these "you are here" signs and Sidney's *Arcadia.*

21. Thus, we could look at the production of "London" and its meaning across gender and class (for example), but we can also look at changes in its meaning across time.

22. Although definitions vary, chorographies can be thought of as descriptions of places through time rather than pictorial renderings. Howard Marchitello makes this distinction between cartography and chorography clear. Maps like Saxton's "work because they seem not to tell stories, seem not to narrate, but, rather, appear to describe objectively the phenomenal world" and "attempt to present their texts as ahistorical"; they are "synchronic representation[s]" in which "history and politics are both subsumed . . . as matters of natural fact,

facts that the map implicitly speaks but refuses to represent explicitly"
(30). Chorographical representations, on the other hand, are di-
achronic and, like Camden's text "seek to represent Britain *in time:*
the land persists, as does its history. . . . Chorography is interpretation
deployed through time; cartography is interpretation ostensibly outside of, or in spite of, time" (30). Nevertheless, Marchitello cautions,
it is important to realize that "maps are narrative, even though cartographic ideology radically denies the narratological nature of the
cartographic practice and its artifacts" (32).

The distinction between *geography* and chorography is made a bit
differently, however, by Claudius Ptolemy in his *Geography:*

> Geography is a representation in picture of the whole known
> world together with the phenomena which are contained
> therein. It differs from Chorography in that Chorography, se-
> lecting certain places from the whole, treats more fully the
> particulars of each by themselves—even dealing with the
> smallest conceivable localities, such as harbors, farms, villages,
> river courses, and such like. . . .
>
> The end of Chorography is to deal separately with a part
> of the whole, as if one were to paint only the eye or the ear
> by itself. The task of Geography is to survey the whole in its
> just proportions, as one would the entire head. (Ptolemy 25)

Perhaps today the distinction might be compared to the difference
between a snapshot and a movie.

23. As late as 1640, in fact, east was sometimes shown at the top of some
    British maps. See Rodney Shirley for additional examples. See De-
    lano-Smith and Kain for the *Totius Britanniae tabula chorographica*
    (21).

24. Consider the title of Thomas Porter's map, which incorporates sev-
    eral such claims: "The Newest and Exactest MAPP of the most Famous
    Cities LONDON and WESTMINSTER with their Suburbs; and the man-
    ner of their streets: With the Names of the Chiefest of them Written
    at Length and Numbers set in the rest in sted of Names. The which
    Names are at Length in the Table with Numbers how to Guide them
    Readily So that it is a ready Helpe or Guide to direct Country-men
    and strangers to finde the nearest way from one place to another by
    T. Porter."

25. The entry appears in the manuscript version of Norden's work but
    not in the subsequent published version.

26. *The Surveiors Dialogue* was printed three times: 1607, 1610, and
    1618. It was expanded with the 1610 edition from five books to six.

27. Elyot also considered the importance of maps for the business of gov-
    ernment and for and for military intelligence. He notes that

by the feat of portraiture or painting a captain may describe
the country of his adversary whereby he shall eschew the dangerous passages with his host or navy; also perceive the places
of advantage, the form of embattling of his enemies, the situation of his camp for his most surety, the strength or weak-
ness of the town or fortress which he intendeth to assault.
(23–4)

28. Along the same lines, Tom Conley suggests, of a portrait of France's
Henry IV on the verso of the title page of some editions of Maurice
Bouguereau's, *Le théâtre françoys,* that "Henry's face does not just
gaze at us. It also seems to look backward, quasi-panoptically, con-
firming that when we look through the folio and the portrait imposed upon the map, the king's mouth is placed at a position that is
identical to that of the crown over the escutcheons of Navarre and
France in front of the arch on the title page" (217). For Conley's
complete discussion of this image, see pages 211–220.

29. The Armada Portrait can be viewed on the Internet at:
<http://www.tudorhistory.org/elizabeth/armadalarge.jpg>. Repre-
sentations of the queen's body on the face of maps, such as an anonymous Dutch engraving of Elizabeth as Europa, may serve the same
purpose (as I discuss in chapter 3).

30. See Burghley's annotated map of Northumberland (British Library,
Royal 18.D.iii.ff 71v–72, in P. D. A. Harvey 1993 *Maps,* 56, Figure
36), his own sketch map of the area and families of Liddesdale, a de-
bated land between England and Scotland (Public Record Office, *SP
59/5, f 44,* shown in Harvey 53, Figure 34), and his annotated map
of the Bristol Channel and genealogical table (British Library, *Lands-
down MS 104, ff 100–101,* in Barber II, 72, Figure 3.7).

31. Machiavelli makes a nearly identical argument in *The Prince* (chapter
14).

32. Beacon towers, used for warning of imminent danger, were also used
for triangulation points in local surveying, and the watchers at these
beacons seem to have helped Christopher Saxton along his way as he
conducted his county surveys beginning in 1567. Two hundred years
later, these beacon points were used for triangulation points for the
Ordnance Survey (Ravenhill 24–25).

33. The highly decorated map of Istelworth Hundred at Syon House by
Moses Glover contains wonderful examples of these features.

34. The most famous is the memory theater of Giulio Camillo (see
Frances Yates 1966, 129–172).

35. Mary Carruthers has "adopted for [this unknown teacher] his Medieval English name of Tully" (see 307, note 116).

36. During Camillo's life, an actual memory theater was constructed in
Italy for the King of France; Viglius Zuichemus wrote to Erasmus

about it in 1532 (Frances Yates 1966, 129–133). Yates has provided a schematic drawing of the theater (insert after p. 144).

37. John Gillies' *Shakespeare and the Geography of Difference* deals with this very subject.

CHAPTER 2

1. This chapter is an expansion of a paper presented in a Spenser Society of America Session at the MLA National Convention, Washington, D.C., December 1996. An earlier version of this paper, entitled "Spenser's Mythic Geographies: Translating the Pictorial Map into the Spenserian Text," was presented at a conference sponsored by the University of Colorado Center for British Studies, "From Robin Hood to Pocahontas: Texts and Their Transmission in the Middle Ages and the Renaissance," November 1996. All quotations from *The Faerie Queene* are from the Longman edition, A. C. Hamilton, ed. Quotations from other Spenser poems are from *The Yale Edition of the Shorter Poems of Edmund Spenser,* ed. William A. Oram, et al., unless otherwise noted.

2. Thus, both John Dee and Francis Drake presented maps of the New World to Queen Elizabeth in 1580 to gain support for English colonization of America. John Dee's map of part of the Northern Hemisphere had a written justification of English imperialism on the reverse, in an attempt to garner more active support for overseas ventures. Likewise, Francis Drake surely had hopes that the world map of the lands he discovered and claimed for England during his 1577–80 circumnavigation would persuade Elizabeth to support planting English colonies in these regions. Although the queen was not persuaded in these cases, by 1620 Drake's map was displayed in the Privy Gallery at Whitehall (Barber II, 66).

3. Royal chambers were frequently decorated with such maps, providing a form of *tableau vivant* in which monarchs made their public appearances. The chamber of Henry III at Westminster had a *mappamundi* on the wall behind the throne, and the audience chamber of Countess Adela of Blois (sister of William the Conqueror) had a *mappamundi* on the floor, scenes from history around the walls, and paintings of the heavens on the ceiling (Barber I, 26).

Katherine Eggert has suggested to me that these displays are a form of centering the monarch in the cosmos, which later turns into the centering of the monarch in courtly society, as Stephen Orgel describes in *The Illusion of Power:*

> Thus the ruler gradually redefines himself through the illusionist's art, from a hero, the center of a court and a culture, to the god of power, the center of a universe. Annually he transforms winter to spring, renders the savage wilderness

benign, makes earth fruitful, restores the golden age. We tend to see in such productions only elegant compliments offered to the monarch. In fact they are offered not to him but by him, and they are direct political assertions. (52)

Similarly, the celebration of the entry of Katherine of Aragon into London in 1501 to marry Henry VII's son Arthur "included a pageant in which, in an effort to compare her betrothed with the sun, a figure representing Arthur, Prince of Wales, was shown enthroned in majesty at the center of the universe" (Barber I, 27). Likewise, Elizabeth is made the center of the universe in *Colin Clouts Come Home Againe* (lines 40–43). Leicester's audacious self-presentation in the Hague in 1586 included a triumphal arch in which Leicester, as "Arthur," is the center of the universe (Frye 1993, 94–95, and Figure 15). I would also contend that the "heaven rooms" and "hell staircases" of some Elizabethan country houses, including William Cecil's Burghley House, provided similar symbolic and ceremonial spaces. The Burghley House website provides a tour that includes such rooms: <http://www.stamford.co.uk/burghley/heaven.htm>.

Some portraits of Elizabeth perform similar cultural work: the frontispiece of George de la Mothe's *Hymne* (ca. 1592) shows her "seated on a globe of the world, between the sun and the moon, with rays emanating from her head" (Strong, *Portraits,* "Drawings & Illustrations," 10); the frontispiece of Saxton's *Atlas of England and Wales* places her portrait between figures of Geography and Cosmography (1579); in the Armada Portrait Elizabeth has her hand on the globe; and in the woodcut from John Cases's *Sphæra Civitatis* (which I discuss in the next chapter) she looms over the universe. As I suggested in chapter 1, her coat of arms on the face of maps, likewise, suggests her dominion over many mapped regions.

4. Ortelius would later suggest in the introduction to his 1606 English edition of his *Theatrum* that a person using his atlas might become "a nobleman in miniature"; he goes on discuss the "discommodities" of rolled or folded maps that require a princely space for viewing or display. His atlas, he claims, solves this problem. (See Conley's discussion 334, n. 11). Ortelius's atlas may not require a princely space, but its size and cost would restrict its readership to a great extent.

5. A brief history of the publication of *The Faerie Queene* might be helpful before beginning. Spenser's first publication of *The Faerie Queene* in 1590 consisted of only the first three books and ended with Amoret "ouercommen quight / Of huge affection" (3.12.45) as she and Scudamour are reunited and "melt" in pleasure:

> Had ye them seene, ye would haue surely thought,
> That they had beene that faire *Hermaphrodite*,
> Which that rich *Romane* of white marble wrought,

And in his costly Bath causd to bee site:
So seemd those two, as growne together quite,
That *Britomart* halfe enuying their blesse,
Was much empassiond in her gentle sprite,
And to her selfe oft wisht like happinesse,
In vaine she wisht, that fate n'ould let her possesse. (3.12.46)

Book 3 in the 1596 edition contained a revised ending that allowed the story to continue. The five final stanzas (43 through 47) of the 1590 edition are replaced with three new stanzas (43 through 45). The 1590 edition included "A Letter of the Authors: To the Right noble, and Valorous, Sir Walter Raleigh knight" (Hamilton *FQ* p. 737–8).

The 1596 *Faerie Queene* included Books 1–6. The *Mutability Cantos* appeared in 1609.

6. I interpret "awe" here as the *OED*'s first meaning: "immediate and active fear, terror, dread."

7. Of course, Spenser's dungeon/map lacks all the details of either map, but the emblem of the ocean as an encircling "river" and of at least one interior river (here, the Styx in 4.11.4) suggests an association with these cartographic paradigms. It is worth reiterating too that not all T-O maps were round, as one would think from the "O" designation (and as shown in Figure 1.1)—many were square, oblong, and oval. For example, the Beatus Map is rather squarish, but nevertheless preserves the T-O design; an image can be viewed at: <http://www.henry-davis.com/MAPS/EMwebpages/207B.html>.

8. See the dragons beneath the world image in Figure 1.3, for example, as well as the monstrous races described in traveler's tales located on the margin of Africa (lower right edge of map). It is not merely the presence of monsters in the margins that facilitates this comparison to the *mappamundi*. Because monsters decorated the margins of maps even in Spenser's day, the monsters alone would not point us to a medieval *mappamundi*, per se. I believe that these marginal decorations do, however, bring us into the realm of cartographic rather than merely decorative art, and these marginal embellishments, coupled with Florimell's position of centrality, facilitate the more specific cartographic association that I am making with medieval maps.

9. See Stephen Booth's discussion of female genitals as "hell" in *Shakespeare's Sonnets,* especially his notes to 119.2, 129.14, and 144.12.

10. By obviating the possibility of rape as abduction, or "secret theft," by Florimell's lovers (in stanza 3), Proteus has Florimell all to himself. The threat of sexual rape, then, comes only from within the "dongeon" rather than from the outside.

11. See Conley (174–201) and Lestringant (1994, 109–110) for further discussion of the genre.

12. Specifically, the entertainments sponsored by Robert Dudley, the Earl of Leicester, at Kenilworth in the summer of 1575 included eighteen days of masques, several of which focus on themes of marriage and one of which centers on a rapist, Sir Bruse sans Pitie. The latter, though cancelled, was included in George Gascoigne's *Princely Pleasures at the Courte at Kenelwoorth. That is to saye, The Copies of all such Verses, Proses, or poetical inventions, and other Devices of Pleasure, as were there deuised, and presented by sundry Gentlemen, before the Quene's Majestie*, published six months after the visit. These entertainments were likewise detailed in *A Letter: Whearin, part of the Entertainment, untoo the Queenz Maiesty, at KILLINGWORTH CASTL, in Warwik Sheer, in this Soomerz Progress, 1575, iz signified: from a freend officer attendant in the Coourt, unto hiz freend a Citizen, and Merchaunt of London*, known as *Laneham's Letter*, published almost immediately after the visit. Susan Frye discusses these entertainments (1993, 56–96). The full texts are found in Nichols (1:420–523).

13. No one, to my knowledge, has done very much with a connection between Florimell and Elizabeth.

Patrick Cheney mentions both Elizabeth and Florimell together in a commentary on Spenser's "virgin wax" reference to the False Florimell (3.8.6) (1997, 55–57 and 285 n. 16).

Elizabeth Bellamy suggests that Elizabeth is ultimately unreadable in the poem; she mentions Florimell three times in her discussion of Elizabeth. Florimell first appears in a parenthetical list of "metonymic displacements of Elizabeth (Belphoebe, Britomart, Florimell, Astraea, Mercilla, etc.)" (8). A footnote to this list develops the correspondence a bit further:

> It is worth mentioning at this point that the metonymic displacement suffered by an unnamed Elizabeth is well illustrated by Arthur's disloyal lamenting for the illusive Florimell: "Oft did he wish, that Lady faire mote bee / His Faery Queene, for whom he did complaine: / Or that his Faery Queene were such, as shee" (3.4.54). The indecisiveness of Arthur, searching for his Faerie Queene but delayed by a longing for Florimell, underscores the interchangeability of the two elusive women. The full identity of each figure is attenuated by a metonymic proximity to the other—neither Gloriana nor Florimell can be wholly herself. (27 n.16)

Florimell appears in Bellamy's essay again as she discusses Ariadne in 6.10, stating, "there is no naming—only the dissolving into analogy that merely supplements, rather than elucidates, and Ariadne takes her place beside Belphoebe, Britomart, Florimell, Astraea—all of *The Faerie Queene's* elliptical but unrevelatory analogues for Elizabeth herself" (17).

Marie Buncombe suggests that Florimell fits Spenser's purposes of fashioning "a gentleman or noble person" as expressed in the letter to Ralegh; moreover she sees Florimell as the avatar of Castiglione's "Neo-Platonic concept of love and the Christian virtues of the chaste, unmarried noblewoman at court" (165), though Buncombe skips 4.11 in her analysis, and never connects Florimell to Elizabeth.

Andrew Hadfield has perhaps done the most, and besides his comments included later in this text, he mentions Florimell in a list of fictional manifestations for the queen, other than Gloriana: "Britomart, Florimell, or even Radigund" (196).

14. Frye provides an insightful discussion of "Spenser's relation to his narrator" in "Of Chastity and Violence" (65–72).

15. All quotations here are from the 1596 version.

16. Foxe's title to his account of Elizabeth's imprisonment also reflects the divine intervention at the heart of his history: "The miraculous preseruation of Lady Elizabeth, nowe Queene of England, from extreme calamity and danger of life, in the time of Queene Mary her sister" (1895). See Frye's discussion of Elizabeth's eventual release (1993, 76–77).

17. Peter Stallybrass describes how "the state, like the virgin was a *hortus conclusus,* an enclosed garden walled off from enemies. In the Ditchley Portrait, Elizabeth I is portrayed standing upon a map of England. As she ushers in the rule of a golden age, she is the imperial virgin, symbolizing, at the same time as she is symbolized by the *hortus conclusus* of the state" (129). In a similar manner, Louis Montrose's discussion of the Armada Portrait connects the body of the monarch with the defeat of the Armada:

> This demure iconography of Elizabeth's virgin-knot suggests a causal relationship between her sanctified chastity and the providential destruction of the Spanish Catholic invaders. . . . The royal body provides an instructive Elizabethan illustration of Mary Douglas's cross-cultural thesis that the body's "boundaries can represent any boundaries which are threatened or precarious." . . . The inviolability of the island realm, the secure boundary of the English nation, is thus made to seem mystically dependent upon the inviolability of the English sovereign, upon the intact condition of the queen's body natural. (1986, 315)

Along similar lines in "Palisading the Elizabethan Body Politic," Linda Woodbridge connects successful bodily invasion during the reign of Elizabeth (in *Titus Andronicus* and *The Rape of Lucrece*) with attempted, but unsuccessful, invasion during James's reign (in *Cymbeline*) when the fear of invasion associated with the female body of the monarch no longer obtained.

But Peter Barber offers another reading of the Ditchley Portrait: he sees Elizabeth's standing on Christopher Saxton's map "as if she were protecting England from storms, which are shown clearing from the west. The armillary spheres she wears as earrings hint at England's imperial future" (II, 78).

18. Oruch provides a useful schematic outline of the inserted fragments and the editions in which they first appeared and a discussion of the poem as a forerunner of Spenser's river marriage (607–608). George Burke Johnston's collection of Camden's poems contains the assembled fragments in Latin as well as two translations: one version is translated by Philemon Holland (88–105), the other by Basil Kennet (137–143).

19. See, for example, a map of "Defences on the lower Thames" drawn by Robert Adams at the time of the Armada and a map of the "Thames estuary" of 1584, which shows areas most vulnerable to attack (figures 33 and 23, respectively; Harvey 1993, *Maps in Tudor England*).

20. See, for example "An Armada Pilot's Survey of the English Coastline, October 1597," in which the pilot describes virtually the entire coastline of England in terms of ease of physical and military invasion and in terms of Catholic reception (Loomis).

21. Ben Jonson's poem "To Penshurst," taken up in chapter 4, celebrates this house.

22. Spenser's praise of Sidney extends from lines 278–315.

23. For a detailed discussion of Spenser's critique of Sidney, see Raphael Falco (95–123)

24. Ironically, Sidney was, on the one hand, protecting the queen's virginity, while on the other hand using assaultive tactics to try to get past her defenses. And Sidney's entertainment *The Four Foster Children of Desire* allegorically stages such an assault. Spenser too is thought to be speaking against the Alençon marriage in the "Aprill" eclogue of *The Shepheardes Calender* and in "Mother Hubberds Tale."

25. "(1566) To a Joint Delegation from Parliament" (qtd. in Frye 1993, 70).

26. Sidney got off rather easily with his banishment from court; Stubbs had a hand amputated, and Wentworth was imprisoned in the Tower until his death.

27. Florimell, too, being set off to one side, is rather "untouchable" (except by Proteus).

28. "A Letter of the Authors: To the Right noble, and Valorous, Sir Walter Raleigh knight" (Hamilton *FQ* p. 737–8).

29. Here, McCoy is actually referring to Sidney's *Lady of May*, presented at Wanstead in 1578, which allowed Elizabeth to have the last word, but the phrase originally referred to the Kenilworth entertainments (McCoy 130; see also 44 and 45).

30. The Garden of Eden is located at the top of the Psalter Map and the Hereford *Mappa Mundi,* for example.

31. According to John Gillies, Ortelius's title and the frontispiece to his *Theatrum Orbis Terrarum* deliberately imitate the Renaissance map room. Gillies considers this frontispiece as the "'*prima foris . . . pagina,*' or 'the first page of the doorway' of a figurative Renaissance maproom of the type which Vasari describes as impressing the visitor with a spectacle of 'all things relating to heaven and earth in one place, without error, so that one could see and measure them together and by themselves.' Such an 'entrance' makes the atlas itself a textual version of the Renaissance genre of the *studiolo,* the *Kunst-und-Wunderkammer,* the monumental *Guardaroba,* or the *Sala del Mappamondo.*" Gillies's discussion here also includes Giovanni Paolo Gallucci's *Theatrum mundi* (Venice 1558), a celestial atlas, and Stanislaw Lubieniecki's *Theatrum cometicum* (Amsterdam 1666–68) (73).

    A recent study, including many photographs and a diagram of the layout of a Sala della Cosmografia, is found in Mary Quinlan-McGrath's "Caprarola's Sala della Cosmografia." Her subject, too, is the wielding of power by use of maps and cosmographic depictions.

    For another excellent discussion of map rooms, including photographs of such rooms, see Juergen Schulz.

32. For the inventories of Leicester House and Kenilworth, see Kingsford; for a discussion of the inventories of maps in royal palaces, see Barber (I, 42–45).

33. A comparison with the contents of Caprarola's Sala della Cosmografia is striking (Quinlan-McGrath).

34. The Orinoco River in Guiana.

35. In Spenser's day, these markers in the actual landscape (many of which are still used today) might have included stakes, stones, crosses, boundary walls, rivers, rocks, and other natural boundaries.

36. To conclude this Proem, Spenser asks pardon and folds (or "enfolds") his map "[i]n couert vele"—he "wrap[s it] in shadowes light" in order to get on with the story of Sir Guyon (2.Proem.5).

37. Or perhaps a "meditation," to use Harry Berger's term (1988, 210).

38. Much of the scholarship on the river marriage episode treats the subject of Spenser's earlier proposed *Epithalamion Thamesis,* a topic worth rehearsing here. Some of the discussion centers around whether Spenser ever wrote this poem (no trace of which has been found) and on how much of the current pageant might be derived from it. In a letter to Gabriel Harvey, Spenser proposed:

> I minde shortely at conuenient leysure, to sette forth a Booke in this kinde, whyche I entitle, *Epithalamion Thamesis,* whych Booke I dare vndertake wil be very profitable for the knowledge, and rare for the Inuention, and manner of handling. For in setting forth the marriage of the Thames: I shewe his

first beginning, and offspring, and all the Countrey, that he passeth thorough, and also describe all the Riuers throughout Englande, whyche came to this Wedding, and their righte names, and right passage, etc. A worke beleeue me, of much labour, wherein nothwithstanding Master *Holinshed* hath muche furthered and aduantaged me, who therein hath bestowed singular paines, in searching oute their firste heades, and sourses: and also in tracing, and dogging oute all their Course, til they fall into the Sea. (Spenser, "Three proper wittie familiar letters," 17)

Berger contends that the current canto is a result of "the revising play of poetic imagination, the mature poet knowing now what he did not then, seeing and using the idea now as he could not then" (1988, 211). Berger suggests, further, that Spenser has moved beyond the map that he sees in the original project—poetry being a higher form of representation (211).

Jack Oruch suggests that "the usual guess about the *Epithalamion Thamesis* is that Spenser did not complete it because he lost interest in the quantitative system, the new 'versifying.' While that may be true, there would have been nothing to prevent him from recasting the poem in an established verse form had he been as enthusiastic about the subject matter as his letter implies. One is inclined to believe that after writing Harvey, Spenser realized the impossibility of treating so large a subject and abandoned it at an early stage" (615; Oruch's full discussion is found on pages 613–624).

Charles Osgood concludes that the present poem could have derived only twenty stanzas from the original poem, if it ever existed (107; his full discussion of sources and influence is found on 100–108). See also Hamilton's note to stanzas 8–53 (*FQ* p. 508).

39. For a detailed history of these maps and of the patronage issues surrounding them, see Helgerson, *Forms of Nationhood* (105–147).

40. In fact, Peter Barber claims that "the Crown's greatest cartographic achievement in the second half of the sixteenth century is to be found in Ireland." Barber points out that in 1550 the English had "little knowledge of Ireland beyond the Dublin Pale. By 1610, by contrast, ministers were familiar with the physical and political geography of the kingdom—in places in considerable detail" (II, 61). Barber goes on to name some of the more significant projects. Cartographers in Ireland were also involved in mapping of plantations, planning against Spanish invasion, keeping down internal revolts, and building fortifications (61–62).

41. J. B. Harley points out that "there are the numerous cases where indigenous place-names of minority groups are suppressed on topographical maps in favour of the standard toponymy of the controlling group" (Maps, 289).

42. According to Barber, the estate map became "the most common form of cartographic activity in Ireland after 1620. English estate surveyors had begun to move across the Irish Channel at the behest of private patrons as early as 1586, however, and some of their work, such as the fine plan of Sir Walter Raleigh's estate of Mogeely, County Cork, in 1598, is the equal of the best then to be found in England" (II, 61–62).

In 1586 four surveyors, Arthur Robins, Francis Jobson, John Lawson and Richard Whittaker, were sent to Ireland to map and measure the newly acquired portions of Counties Waterford, Cork, Limerick and Kerry (Andrews 20).

43. The marriage vows—or marriage ceremony—in fact, seem to occur somewhere *between* stanzas 8 and 9: in stanza 8 we learn that "spousalls" are "[b]etwixt the Medway and the Thames agreed"; stanza 9 tells us that "so both agreed, that this their bridale feast / Should for the Gods in Proteus house be made; / To which they all repayr'd, both most and least"—in this manner, we're "thrust into the middest" of the procession to the feast.

44. Sheila Cavanagh has written that "blaming the victim" is a common reading for rape, and cites several examples of such readings of both Florimell and Amoret (2–3); Camille Paglia finds little to admire in Florimell:

> In *The Faerie Queene*, helpless, retiring femininity is a spiritually deficient persona. Fleeing, ever-receding Florimell, brainwashed by the literary conventions of the love-game, is a caricature of hysterical vulnerability. Terrified by the sound of leaves, she runs even from admirers and rescuers. (184)

Later, Paglia calls Florimell a "professional victim" (186).

45. And perhaps to Spenser, or the narrator, since her rescue must wait until Spenser is good and ready for it. Another example of the representation of "fantasies of defining and controlling" women, as discussed earlier.

46. This subject is taken up in greater detail in chapter 3. An especially useful essay on this subject is Louis Montrose's "The Work of Gender in the Discourse of Discovery."

47. Clare Carroll points out that Spenser's Ireland in *A View* represents the Irish as a "feminized, culturally barbaric, and economically intractable society that must be subjected to complete cultural and economic destruction and reorganization by the English colonists" (163).

48. Montrose, likewise, discusses the association of "Virginia" with the queen's virgin body and the irony that is implicit here (1991).

49. Oruch suggests that "only the 'male' rivers Bregog, Fanchin, and Allo retain their proper titles. Galtymore becomes Arlo Hill with only a slight change in spelling from 'Aherlow,' the name of a neighboring district, but the Mulla River is actually the Awbeg; the Molanna

is the Behanna or Behanagh; Armulla Dale is the valley of the Black-
water or Broadwater; and the parent mountains of Mole are the Bal-
lyhoura and Galtee ranges" (622).

Elsewhere, Spenser writes of two other river marriages—both based
in Ireland. First is the myth of the marriage of the Bregog and the
Mulla, including the story of Old Father Mole's opposition to the marriage and his revenge on the Bregog exacted by throwing stones in his
way and forcing him to dry up (*Colin Clout,* 88–155); second is the
myth of Arlo Hill in which Mulla's sister, Molanna, a nymph of Diana's,
is enticed by Faunas to betray her mistress (*Mutability Cantos,* xi).

For an elaboration of the "founding myth of liberty in the after-
math of sexual violence," see Stephanie Jed's discussion of "Lucretia's rape as a prologue to republican freedom" (5).

50. See Hamilton, *The Faerie Queene,* introductory note to stanzas 8–53,
for several viewpoints (p. 508). In his marriage of the Tame and Isis,
Camden has the Tame as female and the Isis as male. Osgood suggests that the complete reversal that Spenser makes here could be because "Thame is, of course, the smaller stream, and, as we have seen,
in the old poem is the bride of Isis. Spenser may have reversed the relation because Isis is feminine in implication, and because the name
of Thame dominates the new name Thames. And to emphasize this
reversal he has insisted upon the feebleness of Isis, and, contrary to
fact, upon the greater strength of Thame" (72).

51. Consider also her two-stanza description:

> Then came the Bride, the louely *Medua* came,
> Clad in a vesture of vnknowne geare,
> And vncouth fashion, yet her well became;
> That seem'd like siluer, sprinckled here and theare
> With glittering spangs, that did like starres appeare,
> And wau'd vpon, like water Chamelot,
> To hide the metall, which yet euery where
> Bewrayd it selfe, to let men plainely wot,
> It was no mortall worke, that seem'd and yet was not.

> Her goodly lockes adowne her backe did flow
> Vnto her waste, with flowres bescattered,
> The which ambrosiall odours forth did throw
> To all about, and all her shoulders spred
> As a new spring; and likewise on her hed
> A Chapelet of sundry flowers she wore,
> From vnder which the deawy humour shed,
> Did tricle down her haire, like to the hore
> Congealed litle drops, which doe the morne adore.

(4.11.45–46)

52. What she doesn't know is that the poem won't *ever* allow her union with Artegall.

53. Perhaps Elizabeth would consider the figures to be "laughing-stocks," too—another possible reading of "senceles stocks," and certainly a fate Elizabeth would wish to escape. The term was in currency since 1533, according to the *OED*.

54. Joanne Woolway sums up this conundrum well: "The figuring of power through the compass is just one instance in *The Faerie Queene* of the organization of power according to notions of centre and periphery which allow the Queen to dominate the literary space of a poem to which she, although absent from, is nonetheless central. Likewise, on a map, a compass is the centre point from which the geometric organization of the text is projected, and yet, at the same time, a feature which belongs specifically to the map's marginalia" (3).

55. Consider also the commendatory sonnet on the face of the Ditchley Portrait, which reflects the same theme, especially in the last few lines. Roy Strong explains that "at some time the portrait was reduced at the sides, thus mutilating the sonnet in the cartouche to the right" (1964, 75). The parenthetical portions are words and letters interpolated by Strong for the unreadable or cropped off portions of the poem.

> The prince of light, The Sonn by whom thing(s live,)
> Of heauen the glorye, and (of) earthe the g(race,)
> Hath no such glorye as (your) grace to g(ive,)
> Where Correspondencie May haue no place.
> Thunder the Y(m)age of that (po)wer dev(ine,)
> Which all to nothinge with a worde c(
> Is to the earthe, when w(is)dom (f)ayre r(
> Of power the Scepter, not of (   ).
>
> This yle of such both grace (and) power
> The boundles ocean (   ) lus(   ) em(
> P(owerful) p(r)ince (   ) the (   ) ll(
> Riv(er)s of thankes retourne for Springes (shower.)
>
> > Riuers of thankes still to that oc(ean pour,)
> > Where grace is grace aboue, power po(wer.)
> > (Strong 1964, 75–76)

56. And not, lamentably, to the Queen of England. Of course, Elizabeth often used the words "king" and "prince" to refer to herself, but in this context, allying this "King" of the Ocean with the King of Spain makes more sense.

57. Similarly, Tom Conley compares Belleforest's "potamography" of the River Allier in *La cosmographie universelle* and Maurice Bouguereau's attempts in *Le théâtre françoys* (1594); Conley asserts that Bouguereau's "style corresponds to a potamographic map: It pulls the reader into its current, swirls about, crests, and washes over a vast paginal area. Bouguereau attempts to channel or control that flood without sacrificing its content and flow. He deploys rivers to disseminate to the commonwealth the shape of an idealized national geography" (237).

58. So Horace Walpole described her. The Ditchley Portrait and the Armada Portrait are especially striking in this regard.

59. Ralegh himself is frequently associated with the Ocean: in his own *Ocean to Scinthia*, as the Shepherd of the Ocean in *Colin Clouts Come Home Againe*, and in Queen Elizabeth's pet name for him—"Water."

60. Of course, it is not out of character for Spenser to alter topography to fit his own purposes: consider his account of the Thames's change of course in *The Ruines of Time*.

The *Faerie Queene* is a world of myth, and for this world of Spenser's creation, myths like the T-O map, the *mappaemundi*, the Amazon women, and the geographic myths of Ireland fit nicely alongside myths of Arthur, Redcrosse, Britomart, and the rest. John Guillory points out that "the apparent hopelessness of the task set out in these lines is belied by the fact that Spenser has already named (contained) a number of rivers large enough to counterfeit the infinity that he seems to argue lies beyond the power of the artist to represent" (25). Spenser's myth making continues as he describes the task he has just completed (more or less) as being of mythic proportions. Harry Berger suggests that

> One mark of its sophistication, and a major source of its odd panoramic quality, is that the pageant seems to be a meditation on a map of the British Isles. Scudamour's narrative act in IV.x was essentially passive in that he was "copying," that is, remembering, something previously seen. In IV.xi, however, the narrator works not from nature but from an artifact or schema which is itself a triumphant act of mind: the improvements over older methods of cartography introduced during the sixteenth century by Mercator and others produced new and justified respect for this activity in which the mind projected not what the eye saw but rather its own mathematical reorganization of the data. And Spenser does not simply "copy" a map. Though his visualization in effect asks us to be aware of this model, it leaves maps far behind; the principle of meaning and visualization is rhetorical, not cartographic. (1988, 210–211)

I would suggest that the "principle of meaning and visualization" is allegorical *and* cartographic.

When Michael Drayton tries his hand at chorographical poetry in 1625, with his *Poly-Olbion*, he too presents a river-marriage pageant. Complete with illustrations, Drayton's pageant is at once both more fanciful and more factual than Spenser's. It is more fanciful because of the many personified illustrations of geographic features and the simultaneous depictions of various stages of the procession, and it is more factual because rivers of the world do not attend this wedding. As Helgerson says, "his rivers stay in their beds" (1992, 142) and Drayton has even fixed them on the appropriate maps. On the other hand, Spenser shows that by means of a pageant of several maps the great rivers of the world *can* be brought to England.

61. The early modern reader may not have expected the kind of fidelity to geography, especially in allegory, that Helgerson contemplates. Roy Strong suggests, in his discussion of the painting "Queen Elizabeth Going in Procession to Blackfriars in 1600," that

> there is no understanding or use of scale in relation to optical distance, or of the placing of characters and objects within a homogeneous geometric totality governed by the new laws of perspective. . . . There is no notion that a picture's surface should encapsulate a given viewpoint at a single moment in time. (1977, 43)

The procession painting that Strong discusses, in fact, brings together three houses (two in the background and one in the foreground) that may belong together symbolically (as three homes owned by the Worcester family) but not geographically as they are spread out across the English map from the Strand to the Welsh Marches. Similarly, in the Armada Portrait, the arrival of the Armada is shown just behind Elizabeth through one window, while its defeat is shown behind her in an adjacent window. That these events cannot possibly have happened together prompts Strong to conclude that "the idea of a picture being direct reportage of a single event was utterly alien to the Elizabethan mind" (46). In the cartographic realm, too, battles that occurred centuries apart but are of great import as defining moments for particular locales are often depicted together on the face of maps. John Speed's map "The Invasions of England and Ireland with al their Civill Wars since the Conquest" is one example (from *A Prospect of the Most Famous Parts of the World*, 1627 [Plate 70, Shirley]).

62. Roche remarks that "The procession has been compared to the masque because of its pageant-like descriptions, but the effect is not that of a masque—it can hardly be called a procession, for the main effect is that of a static picture. The participants do not move. The poet describing

the procession moves through his description in the same way that one 'reads' a medieval painting, moving from scene to scene within the frame of one picture" (179)—or, I would argue, the way one views a map or a series of maps in a map room—a "theater of the world."

63. Some of the maps listed in inventories of 1547–49 are known to have still been on display in the reign of James I, according to the journal of Johan Ernst, duke of Saxe-Weimar (1613, Barber I, 55, n. 141).

64. Another interesting use of maps for propaganda is that made by Edward Seymour, who, as Lord Somerset and regent to Edward VI, "used maps and plans to exploit his advantage politically and psychologically as well as militarily." Barber tells how in 1548 "he chose to display a now lost plat of England's impressive new fortification of Haddington in Scotland prominently on the wall of the chamber in which he was entertaining the French ambassador." This display and polite warning did not go unnoticed by the French ambassador, but the French King Henri seems to have appropriated this type of display to his own use two years later when his "triumphal entry into Rouen was accompanied by banners containing landscapes of the Scottish lowlands, including Haddington, which had been won back by French arms from the English" (I, 40).

CHAPTER 3

1. An earlier version of part of this chapter appears as "A Room Not One's Own: Feminine Geography in *Cymbeline*," in *Playing the Globe: Genre and Geography in English Renaissance Drama*, ed. John Gillies and Virginia Mason Vaughan (Madison and Teaneck: Fairleigh Dickinson University Press, 1998), pp. 63–85. I wish to thank the publisher for allowing me to republish it here. All quotations from Shakespeare's works are from *The Riverside Shakespeare*, G. Blakemore Evans, ed., unless otherwise noted.

2. Catherine Stimpson uses this evocative title for her discussion of rape in *Titus Andronicus, The Rape of Lucrece,* and *Cymbeline*.

3. Louis Montrose offers an excellent discussion of this engraving in "The Work of Gender in the Discourse of Discovery"; Tom Conley, who also discusses this image, concludes:

> Fathoming the illusion he takes to be America, [Vespucci] focuses on the inner side of his pupils, aiming his gaze at the mirage of his own imagination. Where he finds his origin in the virtual feminization of himself, in the female he believes arising from his own cartographic hammock, his map is surely self-made. (309)

4. Indeed, in *Cymbeline*, England is even "clothed"—in a "salt-water girdle" (3.1.80).

5. Thus invoking but transsexualizing Leonardo's *Man in a Circle and a Square,* discussed in chapter 1.

6. Using the popular iconography of the queen along with other emblems of chastity, Peter Stallybrass discusses women's chastity as a "patriarchal territory" that must be safeguarded. Taking as his model Mikhail Bakhtin's distinction between the classical ("finished, completed") body and the grotesque body (which is "unfinished, outgrows itself, transgresses its own limits" [Bakhtin 26]), Stallybrass shows how women, as possessions of men, were subject to constant surveillance of three specific areas: the mouth, chastity, and the threshold of the house. These three areas are often collapsed into each other so that linguistic fullness and/or frequenting public space might be associated with wantonness (123–142).

7. Tom Conley presents a similar discussion of Maurice Bouguereau's acrostic sonnet on the letters of the name of Henry IV of France: Henry de Bovrbon (or Henri de Bourbon), found on the verso of the title page of Bouguereau's *Le théâtre françoys* (208–220).

8. Similarly, in Donne's poem, "Hymn to God My God, in My Sickness," the poetic persona describes his own body as a map being inspected by his doctors as he awaits death:

> Whilst my physicians by their love are grown
>   Cosmographers, and I their map, who lie
> Flat on this bed, that by them may be shown
>   That this is my south-west discovery
>   *Per fretum febris,* by these straits to die,
> I joy that in these straits, I see my west;
>   For, though their currents yield return to none,
> What shall my west hurt me? As west and east
>   In all flat maps (and I am one) are one,
>   So death doth touch the resurrection. (6–15)

Howard Marchitello presents an excellent discussion of this poem, suggesting that Donne has transformed the conceit of man as microcosm "through the invocation of maps: Donne does not claim to be the world but, rather, its map—not the world but its representation. By distinguishing between being and representation, Donne signals the fundamental act of signification by way of which maps are understood as representations of the world." He goes on to suggest that Donne had in mind "something like Mercator's famous global map"—still "fundamentally heretical" in Donne's day (15).

Lisa Gorton suggests of some of Donne's other poems that "he was fascinated by new discoveries. He took up the modern idiom of maps and discovery with delight. But he was also deeply attached to the past, and his assumptions about space belonged to an old tradition: a cosmographic rather than cartographic way of imagining space." While that may be true in other poems, in this poem and in "Love's Progress," he seems to use the cartographic and cosmographic registers with equal dexterity.

9. Vickers (95–115). Patricia Parker also ties together blazon, inventory, and exploration in *Literary Fat Ladies: Rhetoric, Gender, Property,* Chapter 7, "Rhetorics of Property: Exploration, Inventory, Blazon" (126–154). Linda Woodbridge, too, discusses the land-as-woman trope in "Palisading the Elizabethan Body Politic."

10. In addition to the references in *Lucrece,* Shakespeare also connects woman and geography in *The Merry Wives of Windsor,* where Falstaff describes Mistress Page as "a region in Guiana, all gold and bounty" (1.3.69) and where he articulates his plans to seduce both Mistress Page and Mistress Ford: "They shall be my East and West Indies, and I will trade to them both" (71–2). In *The Comedy of Errors,* Dromio of Syracuse orates a lengthy and very bawdy description of his love Luce as "spherical, like a globe" wherein, he swears, he could "find out countries" (3.2.112–143).

11. An allusion to celestial mapping, too, perhaps. See Glenn Clark's alternate reading of mapping metaphors in *Cymbeline.*

12. The image is compellingly reminiscent of Van der Straet's engraving of Vespucci surveying the recumbent "America." Compare, too, Troilus's description of Cressida:

> Her bed is India, there she lies, a pearl;
> Between our Ilium and where she [resides],
> Let it be call'd the wild and wand'ring flood,
> Ourself the merchant, and this sailing Pandar
> Our doubtful hope, our convoy, and our bark.
> (*Troilus and Cressida* 1.1.100–104)

13. Although the stage direction between lines 24 and 25 was not added until the 1623 Folio, the use of the word "table" here nevertheless warrants discussion in conjunction with the sort of triangulation that Jachimo seems to be configuring. Evidence from the *OED* invites the multiple readings of "table" that inform my interpretation of this key scene. Naturally, a "table" can be a writing tablet (*OED* definition 2), but it is also "a board on which a picture is painted; hence the picture itself" (3), suggesting a pictorial representation. It can also be a geographical table, a map, a chart such as Camden's *Britannia: A Chorographicall table or mappe of Britaine* (10d), or a "plane table" or "plain table" (17g)—the table that a surveyor uses to measure angles and thence to calculate distances by triangulation. Finally, "table" can refer to mathematical tables, such as those used in geographical surveying to calculate area. It would strain plausibility to

think that Jachimo would actually set up a plane table here but the metaphoric association is available nevertheless.

14. As a surveyor or mapmaker would render a sketch for an engraver. Details on a sketch map need not be entirely accurate either. On Saxton's county maps, for example, hills bear no verisimilitude to actual hills in the terrain. Instead, clusters of rather uniform hills are stamped in to represent hilly areas.

15. Homophony and puns are very effective memory cues, according to Carruthers (105).

16. Again, for a discussion of "hell" as female genitalia, see Stephen Booth's explication of *The Sonnets*. Jachimo could also be posing an opposition between *this* heavenly angel and the "hell" beneath the stage, to which he will now return via the trunk.

17. "Reap" and "rape" could also be homonymic in the pronunciation of the day. The bracelet, too, like Nerissa's "ring" at the end of *The Merchant of Venice* (5.1.307), has sexual implications.

18. Compare also Jachimo's initial impression of Imogen, filled with oxymorons, the frequent refuge for Shakespearean characters under stress:

> JACHIMO: Sluttery, to such neat excellence oppos'd,
>   Should make desire vomit emptiness,
>   Not so allur'd to feed.
> IMOGEN: What is the matter, trow?
> JACHIMO: The cloyed will—
>   That satiate yet unsatisfied desire, that tub
>   Both fill'd and running—ravening first the lamb,
>   Longs after for the garbage.
> IMOGEN: What, dear sir,
>   Thus raps you? Are you well?
>
> (1.6.44–51)

19. Of course in his "woman's part" speech in the next scene, Posthumus admits to a rather disappointing connubial union in which Imogen "restrain'd" him of his "lawful pleasures" and "pray'd me oft forbearance" (2.5.9–10). In some texts the "woman's part" speech is at 2.4.153*ff*, following the scene divisions of the First Folio (The Arden Shakespeare *Cymbeline* is one example [Nosworthy]).

20. From the Greek *autopsis: aut-* + *opsis*, sight. The word did not take on its most common modern meaning of the dissection of a corpse until later in the seventeenth century. But that later meaning is certainly tempting in a scene where the "ape of death" (2.2.31) lies over the woman being inspected.

21. See Hartog, "The Eye and the Ear," Chapter 7 in *The Mirror of Herodotus,* for a discussion of this hierarchy of evidence (260–309).

22. See Stephen Greenblatt, *Marvelous Possessions: The Wonder of the New World,* for a thorough discussion of marvels and wonders of the New World.

23. The men have indeed chosen each other as rivals first and foremost; their rivalry defines their relationship before Jachimo ever *sees* Imogen.

24. This would be high praise, indeed; according to Carruthers,

> the trained memory was not considered to be merely practical "know-how," a useful gimmick that one might indulge in or not (rather like buying better software). It was co-extensive with wisdom and knowledge, but it was more—as a condition of prudence, possessing a well-trained memory was morally virtuous in itself. The medieval regard for memory always has this moral force to it, analogous to the high moral power which the Romantics were later to accord to the imagination, genetrix of what is best in human nature. (71)

Of course, the "moral force" assumed to underlie a memory like Jachimo's and a compliment like Posthumus's is deficient in this case; still, Posthumus does not know what a cad Jachimo is, and it seems here that Jachimo's good memory is all the recommendation he (and his account of events) needs to be considered reliable.

25. Noting and remembering the circumstances under which one created a particular memory image is considered an asset to artificial memory systems, too, as Hugh of St. Victor advocates:

> We [should] also pay attention carefully to those circumstances of things which can occur accidentally and externally, so that for example, we recall along with the appearance or character or location of the places in which we heard one thing or another, the face and habit of the people from whom we learned this and that, insofar as they are the kind by which they accompany their performance of an activity. (Qtd. in Carruthers 95)

26. Patricia Parker's discussion of *enargeia* in "Shakespeare and rhetoric: 'dilation' and 'delation' in *Othello*," is germane to this point as well: Iago is thereby able to describe Desdemona's supposed infidelity so vividly that Othello accepts a mere description as "ocular proof" and responds with, "I see" (*Othello,* 3.3.360*ff*).

27. J. M. Nosworthy suggests that "Imogen," while used throughout the 1623 Folio, may be a misprint for "Innogen" (or Jnnogen), the name Simon Forman used in his diary when describing the 1611 staging of the play. Nosworthy also points out the occurrence of a mute Innogen, wife of Leonato, in the 1600 Quarto of *Much Ado About Nothing,* demonstrating a long-standing association of the names of this hero and heroine, since Posthumus's family name is

Leonatus (7). Also important, I think, is this Innogen's status as a mute or "ghost" character in *Much Ado,* similar to Imogen's diminished political status by the end of *Cymbeline.* The Norton and Oxford editions of the play use the name Innogen.

28. Garrett Sullivan's work on maps and roads in Wales is quite useful here ("Civilizing Wales," and *The Drama of Landscape* [127–158]).

29. As we noticed earlier, the mole becomes a "stain" only *after* the tale is told.

30. The difference in treatment of the antagonist is clear from the following: "Than toke the officers John of Florence and brought hym besyde of the galowes, where the Justice should be done. And whan that he had made his prayers and all doone, than made the hangman him knele downe and smote Johans of Florence head of[f], and after that laied his body vpon a whele, and the head he stycked on a stake and set it by, ouer the head a galowes: all after the maner as the kyng had iudged him; and than retourned home againe. And in this maner was John of Florence serued for his great falshed and thefte that he hadde done to that trewe wyfe and mayde" (*Frederyke of Jennen,* reprinted in Nosworthy, 203).

31. Just as John Davies suggested that poetry is a better means of portraiture than painting, Shakespeare promotes the tale as better than the map.

32. This is one of the main streams of John Gillies's argument in *Shakespeare and the Geography of Difference.*

CHAPTER 4

1. This idea is qualified by William McClung, however, who sees this same contrast of "good" (old) wealth vs. "new wealth" in the Roman poems.

2. Besides referring to the labor necessary to build the walls, the "ruine" and "grone" could also refer to enclosure, as Williams acknowledges later. In the case of Penshurst, that enclosure would have taken place before the Sidneys acquired the property, so any neighbor-ruin that might have followed is not of their making.

3. Don Wayne briefly suggests the possibility of rent-in-kind but seems inclined to dismiss such a possibility by summarily turning to a more conventional reading centered around what he calls a "chain of giving" at Penshurst (75). See also McClung (119–120) and Alastair Fowler's refutation of such a suggestion (1964, 111).

4. One of Agas's advertisements that has survived gives a sense of his dedication to the profession and the skills one must have to call oneself a surveyor:

> To all persons whom these presents may concerne,
> *of what estate and degree soever.*

No man may arrogate to himselfe the name and title of a perfect and absolute Surueior of Castels, Manners, Lands, and Tenements, unlesse he be able in true forme, measure, quantitie, and proportion, to plat the same in their particulars, ad infinitum, and thereupon to retrive, and beat out all decaied, concealed, and hidden parcels thereof, fitting the same to their euidence, how ancient soever; although blemished, obliterate, and very much worne: besides the quickening and reuiuing of Rents, Customes, Liberties, Priuileges, &c. thereunto belonging: with perfect knowlege of customarie Tenures and Titles of all sorts: framing Entries accordingly: together with good and commendable penmanship, as well for the Plat, as Booke, from the same. And for that more abuse in concealments incroachments, &c. hath beene offered in these last 100. yeeres, than in 500 before, and that many doe now refuse (as more heereafter will) to pay their rents and duties, otherwise than on the meeres head (their Lands and Tenements first singled out, and set foorth vnto them, *metis & bundis*) I may not terme him so much as a Surveyer, that performeth not these difficulties, and such like incident to Suruey.

By Radulph Agas of Stoke next Nayland in Suffolck.
Practised in Survey more than 40. yeeres. (*STC* 195.5)

5. In semiotics, of course, the best sign of an object *is* the object itself, so the farmer isn't too far afield with his suggestion.
6. From "*any* chair," as Garrett Sullivan asserts (*Drama* 43).
7. This rather odd commonplace is also reflected in Robert Herrick's "The Country Life": "yet thou dost know, / That the best compost for the Lands / Is the wise Masters Feet, and Hands" (22–24).
8. We might think of Bottom and his league of rude mechanicals in *A Midsummer Night's Dream*. While there is no evidence that they lack competence in their own crafts, they are certainly miserably inept when they venture into other crafts—such as theater.
9. Norden does outline the techniques of measurement to the Farmer, but Norden's instructions, though informative, do not constitute anything like a technical guide to performing a survey—for that kind of instruction we would need to go to another source, such as Aaron Rathborne's work. Norden seems to include this rough outline of scientific methodology primarily as a way of legitimizing his own craft.
10. Some of the forerunners of pictorial renderings of the countryside of England had titles that emphasized their peripatetic underpinnings: William Lambarde's *A Perambulation of Kent* (1576) and *The Itinerary* of John Leland (1535–1543) are two examples.

11. Today, we can view Penshurst via the Internet at the official website: <http://www.penshurstplace.com>, as well as an unofficial one: <http://www.i-way.co.uk/~sid/penshurst.html>.

12. This could be why Jonson doesn't describe the gardens in much detail.

13. Thomas Littleton's (or Lyttleton's) *Tenures,* a book on tenure law, was so widely read that it was reprinted at least forty-six times in French and thirty-four times in English between 1482 and 1630 and was the basis of Edmund Coke's monumental work on English land law. The number of editions plus the variety of formats in which it was published (folios, quartos, octavos, duodecimos, and sexidecimos) attests to the work's widespread popularity with various groups of readers. The larger, more expensive folios could be spread out on a desk to be used by lords, lawyers, and scholars, while quartos, octavos, and the other formats were smaller, cheaper (and hence more available to less cultivated readers), and could be carried in a pocket for quick reference. For a discussion of the distribution and use of the various sizes of printed books, see Arthur F. Marotti (286–290). For a list of all of the editions of Littleton, see *STC* entries 15719–15783.

14. Leigh enumerates many kinds of mills in this category: "corne Milles, Horsmilles, watermilles, windemilles, or Quarnes for graine, or other milles, as Smith milles, Iron milles, fulling milles, Sythmilles, Cutler milles, Tynne blast milles, Lead blast milles, or suche like" (Leigh C4ᵛ).

15. Raphael Falco comes closest to this reading in his suggestion regarding all the types of woods being brought indoors and used for warmth, as the title to his chapter on Jonson suggests, "A Fire Now, That Lent a Shade" (124–165).

16. See Alastair Fowler's discussion of "Georgic and Pastoral."

17. According to Leigh, the building of some walls could actually trigger an "encrease of rent": "where the lorde hath suffered his Tenauntes, or any of them to builde, or set any wall, Barne, Stall, or House, upon any parte of the lordes Waste, or Soile, or that the Lorde hath granted any Tenaunte to enclose any part of his common, or Waste grounde, or suche like, and reserueth to hymself an yerely Rente for the same, and suche like" (Leigh D3ʳ⁻ᵛ).

18. The paying of rent and "hav[ing] no sute" are not mutually exclusive.

19. The two traditional feasts at which rent was paid were Lady Day (or Annunciation, March 25) and Michaelmas Day (September 29). Roy Christian explains that rent days could be days of great festival (65).

20. "Brueghel" is also frequently spelled "Breughel." A color image is available on the Burghley House Web site at: <http://www.stamford.co.uk/burghley/tour1.htm>.

21. The actual "offerings" vary from one version of the painting to another (and at least thirty-seven versions of this painting are known to exist, dated from 1615–21); see Georges Marlier, *Pierre Brueghel Le Jeune* (435–440).

22. Even though the Sidneys are a family who, despite being infinitely better off than the farmers themselves, are nevertheless widely known to be "cash poor."

23. For Rogation Sunday, the fifth Sunday after Easter, the Psalm is 84: "O howe amiable are thy dwellinges, Thou Lorde of Hostes?"

24. Rogation celebrations occurred all over Europe. Martin Luther wrote a meditation for Rogation Week ("Von dem Gebeet und Procession yn der Creützwochen" [On prayer and procession in Rogation Week], 1519) and Bach's Cantatas 86 and 87 were originally written for the occasion. Rogation celebrations are still widely observed in various parishes, manors, and villages in England. Many parishes within London still beat their bounds, beat the waters, and beat the boys; at the Tower of London thirty-one markers are beaten. Some villages beat their bounds on other ecclesiastical holidays, such as the Feast of the Blessed Virgin Mary and St. Bartholomew's Day (see Christian [56–67] and Cooper and Sullivan's entry for "Rogationtide" and "24 August" for examples).

25. Modern surveys, title searches, and title insurance policies perform similar work today.

26. Garrett Sullivan discusses a similar paradigm shift that made the plane table, which did not require mathematical knowledge, a subject of derision among those adept at mathematical calculations. Among its detractors was Thomas Digges, who in 1616 called it "an Instrument onely for the ignorante and unlearned, that have no knowledge of Noumbers" (qtd. in Sullivan, *Drama* 40). The theodolite is the "badge" of the more learned—one who is "a master of a set of practices inaccessible to most," according to Sullivan (41). His discussion of the arguments regarding the plane table and theodolite is found on 39–41.

27. The manor house was built by a London merchant and financier, Sir John de Poulteney, in 1338–49, and centered on the medieval great hall. It was fortified later in the fourteenth century by John Devereux. After it was given to the Sidneys by Edward VI in 1552, several additions were made.

28. See Richard Helgerson, *Self-Crowned Laureates* (101–184) and Raphael Falco, *Conceived Presences* (124–165) for informative discussions of Jonson's poetic ambitions.

29. Once again, Peter Stallybrass's "Patriarchal Territories" is useful. Malcolm Kelsall's work is useful here, too, in reminding us that this wall, while not part of a castle keep, or fortification, bears the reminder of the fortified castle or home (and Penshurst was fortified at one time, 40–41). The walls also bear the reminder of the sort of military siege that we are encouraged to forget in this poetic interlude of tranquility. Notably, in the poem, Sir Philip Sidney is remembered as a poet—his birth attended by muses; his death in voluntary battle is ignored, as is his vexed relationship with Queen Elizabeth.

30. Again reminding us of Agrippa's remark of Cleopatra: "She made great Caesar lay his sword to bed; / He plough'd her and she cropp'd" (*Antony and Cleopatra* 2.2.227–8).

31. See Partridge, "plum" (163–4) and "medlar" (147). See also Frankie Rubenstein, "plum" (198) and "pear" (189–190); Rubenstein's first meaning for both words is "testicles," with additional implications dealing with male genitals and copulation ("pear" and "pair"). One wonders when the tradition of farmer's daughter's jokes began, if here the daughters are carrying emblems that signify or advertise their sexual availability and appetite, symbolizing not only their own genitals but also signifying male genitals.

32. And it is their fathers who control their sexuality—they are the ones, after all, who "commend" them "*this way* to husbands"—meaning either "in this direction" (that is, "*to* Penshurst") or in this manner (that is, offering themselves like a commodity—fruit). The daughters are clearly a "patriarchal territory."

33. Although these possessions could be the lady's dower, their status here, like hers, is in service to the lord.

34. Cuckoldry has a certain edginess that seems hard for authors of the early modern period to resist. Shakespeare, too, seems never to be able to resist the barb:

> VINCENTIO. Art thou his father?
> PEDANT.   Ay, sir; so his mother says, if I may believe her.
> (*The Taming of the Shrew* 5.1.32–34)

See also *The Tempest* and *King Lear*.

35. Garrett Sullivan provides a useful discussion of one woman who is notoriously disloyal to her husband and thus to the land in his discussion of Alice Arden in the anonymous play *Arden of Faversham*:

> The point is not merely that Arden sees Alice as a piece of property who might be stolen by Mosby, but that Arden's relationship to the land is intertwined with his relationship with his wife. Mosby constitutes a threat to the passing down of Arden's property to his heirs, and Alice's love for Mosby cannot be separated from the potential mismanagement of her husband's affairs. . . . Alice is seen as jeopardizing what "belongs" to her husband. (*Drama* 49)

36. There is also a quibble in "housewifery" and "hussy" that is useful here; the lady's absence on the occasion of the king's visit may hint at this paronomastic association and the actual threat to the lady's chastity that her being away may exacerbate. It is certainly not beyond Jonson's sardonic humor to make such a hint. Nor is it beyond him, I would conjecture, to make an oxymoronic pun five lines later

when he says that the lady is "chaste withal"—or chaste *with all* (one can hardly be both "chaste" and "with all").

37. An action frequently lampooned in the city comedies, a subject taken up in my last chapter. "With time," James I declared in 1616, "England will onely be London, and the whole countrey be left waste." (speech in Star Chamber, June 20, 1616, in C. H. McIlwain, ed., *The Political Works of James I* [Cambridge, Mass.: Harvard University Press, 1918] p. 343, qtd. in Manley, *Literature and Culture*, 133).

38. See especially Falco (146–47) for a discussion of the poetic genealogy that Jonson establishes for himself.

CHAPTER 5

1. I am not alone in this reading of Mullaney. Garrett Sullivan voices similar misgivings, saying: "Mullaney is persuasive in his discussion of the importance of ritual to early modern London, but his account goes too far, in that he sees little other than ritual or its traces in the spaces of the city. . . . While, as Mullaney states, ceremony is a way in which the city defines itself, it is not the only way, and we must understand that much of what the city contains—for example, foreign-born workers, nonguild industry, and a seasonally employed work force—is not registered within its rituals" (*Drama* 201). Sullivan finds one solution in the application of the theories of Henri Lefebvre, "in the way in which spatial practice contradicts a notion of the wall as representational space: the wall is no longer 'an effective barrier against the outside world' because the socioeconomic expansion of the city means that outside cannot be easily separated from inside" (206).

2. And once we know this joke, we reread the Induction with newfound knowledge as "east" and "west" debate their own merits (Prologue 1–14).

3. See especially *Michaelmas Term* and *The Roaring Girl*.

4. Although as we have seen, that praise may be mixed with a dose of irony.

5. Mullaney's reading tries to create too easy a dichotomy between the city and the outskirts, though perhaps the city fathers would like to think that all the misfits are successfully ushered outside the city's limits. There are, of course, contradictions within the city itself, and it is not only ceremony that defines the walled portion of the city. Mullaney is correct, though, in stating that the contradictions of the city were played out on the stage in the Liberties.

    Foucault coined the term "heterotopia" to describe places in society that don't seem to fit, but are nevertheless present—sometimes in the margins and sometimes in the center. Examples he gives include the cemetery, the boarding school, the honeymoon trip, museums, libraries, and fairgrounds. An important distinction of heterotopias is that they

> presuppose a system of opening and closing that both isolates them and makes them penetrable. In general, the heterotopic site is not freely accessible like a public place. Either the entry is compulsory, as in the case of entering a barracks or a prison, or else the individual has to submit to rites and purifications. (1986, 26)

6. The word "panorama" (Greek for "all-embracing view") was not coined until 1789 when it was first used to describe Robert Barker's 360° view of Edinburgh (Hyde 11).

7. This is a popular model for tourist maps even today.

8. Two plates of the Copperplate Map can be viewed online at the Corporation of London's *Collage* website: <http://collage.nhil.com>; enter "Copperplate Map" as the search term.

9. Harvey points out that "the survival of very large maps on paper is bound to be chancy" (*Maps in Tudor England*, 74); and Ralph Hyde adds that both maps and prospects were hung on walls unframed: "within twenty years, one can well imagine, dirty and torn, they would be thrown away" (Hyde 14). Only one copy survives of the anonymous "View of the Cittye of London from the North Towards the Sowth" (ca. 1596), having been bound in a manuscript journal. Only one complete copy and one incomplete copy survive of Norden's *Civitas Londini*.

10. These three existing plates, recovered at various dates in the twentieth century, were found on the backs of late sixteenth-century and early seventeenth-century Flemish paintings. One was found in 1955, another in 1962, and the third in 1998. The paintings on the reverse of first two plates are *The Tower of Babel* (ca. 1595), by the circle of Martin Van Valkenborch, and *The Assumption and Coronation of the Virgin* (ca. 1620), attributed to Frans Franken. The third plate, from the Dessau Art Gallery in Germany, features another Van Valkenborch circle *Tower of Babel* (ca. 1600). Each of these plates is displayed at the Museum of London so that both the painting and the engraving may be viewed.

11. This map was almost certainly *not* drawn by Ralph Agas, though this moniker has now become so well-known that the map is still associated with his name (See Marks for a discussion of the authorship of this map, 21). The map can be viewed online at the *Collage* website: <http://collage.nhil.com>; enter "Agas woodcut" as the search term.

12. This map was certainly drawn much earlier, however. At least four major changes to the face of London during the time period help in

dating early maps of London and lead to the conclusion that Braun and Hogenberg's map was most likely drawn in the 1550s. These four changes are: the house known as Suffolk Place was purchased by the Archbishop of York in 1557 and became known as York Place; the cross at St. Bardolf without Bishopsgate was destroyed in 1559; St Paul's spire was destroyed by lightning in 1561; and the Royal Exchange was built in 1566–1570 and opened by Queen Elizabeth in January 1571. Interestingly, between the first state of their map (1572 and 1574) and the second state (1575 and later), Braun and Hogenberg account only for the addition of the Royal Exchange, attesting to its importance in declaring London a center of international trade. They leave St. Paul's spire intact—a landmark that must have been very important in the civic and cultural memory, but also about whose destruction Braun and Hogenberg might not have known since they were foreigners; the fame of the newly opened Exchange, however, had apparently spread.

Both a black and white and a colored print can be viewed online at the *Collage* website: <http://collage.nhil.com>; enter "Hogenberg" as the search term.

13. Published between 1572–1617, this six-volume work depicts the major cities of the early modern period, 531 cities in all. Interestingly, the original rationale for populating a map with human figures was that doing so would prevent their examination by the Turks, whose religion forbade them to look at representations of the human form (Elliot 28). But what began as a kind of "defense mechanism" became a widespread artistic convention—many maps of this period depicted local citizens, in native costume, engaged in daily activities.

14. A translation of the cartouche in the lower left corner of the map helps to underscore the importance of commerce:

> This is that royal city of all England London, situated on the River Thames, named, as many believe, by Caesar after the Trinobantes, made noble by the commerce of many races, adorned with houses, decorated with temples, elevated with arches, with famous ingenuity/artifice, with its men teaching all skills/arts, who are generally outstanding and renowned; finally its wonderful excellence of wealth and abundance of all things; this same Thames carries into it the wealth of the whole world, a passageway for laden ships for sixty miles, navigable to the city with a very deep hull. (Translation courtesy of Matthew Payne and Stephen Freeth of the Guildhall Library, Manuscripts Department)

Peter Barber suggests that a Hanseatic connection would help to explain the content of the right cartouche, which praises the Hanseatic league (of which London was not even a part) and touches on the significance of London only in the last sentence. The words may have been poorly copied from a missing panel of the Copperplate Map (Barber 2001). Although their appeal seems to be aimed at Braun and Hogenberg's continental (and home) audience, it is significant to note that the subject matter is once again commerce:

> Stilliaryds) The Hanse, a German word, meaning an assembly or community, is a confederate Society of many cities, instituted, firstly on account of loans from Kings and benefices from Dukes, and then for the safe management of merchandise by land and sea, and finally to conserve the tranquil peace of the Republic and the modest education of the youth; it was adorned with privileges and immunity from the taxes of very many a king and prince, mostly of England, France, Denmark and Great Muscovy, and especially the Duke of Flanders and Brabant; it has four markets, which some call *Cuntores,* in which the businessmen of the cities reside and practice their business; one of these, here in London, the domestic economy flourishes, and has as its base a Teutonic Guildhall, which is commonly known as the Stiliard. (Translation courtesy of Matthew Payne and Stephen Freeth of the Guildhall Library, Manuscripts Department)

15. Of particular note are those of Oxford and Cambridge. The map of Oxford, known from a single copy, was printed from eight plates, measured three feet by four feet, and was drawn by Ralph Agas in 1578. The Cambridge map was printed from ten plates, drawn by an unknown mapmaker, but engraved by Richard Lyne and published in 1574. A plan of Exeter followed in 1587, and by 1600 Braun and Hogenberg's collection of "Cities of the World" included plans of Bristol, Canterbury, Chester, and Norwich, as well as London.

    Some colorful manuscript maps of several towns that have survived also attest to the civic pride of the time. Of special note are a manuscript of Shrewsbury from the late sixteenth century, preserved in Lord Burghley's private atlas (British Library, Royal MS. 18.D.iii, ff.89v–90 [figure 36 in P. D. A. Harvey, *Maps in Tudor England*]) and a map of Norwich by William Smith, 1588 (BL, Sloane MS 2596, f.61 [figure 52 in Harvey]).

16. Christopher Saxton's earlier *Counties of England and Wales* (1567) did not include such plans.

17. Several of Norden's maps can be viewed at the Corporation of London's *Collage* website: <http://collage.nhil.com>; enter the search term "John Norden."

18. Ralph Hyde summarizes R. A. Skelton's classification of town "views" into four types: first is the "stereographic views or profile views . . . [in which] the town presented itself to the eye of an ob-

server at a point on the ground or not far above it"; second is the "prospective views, birds-eye views" (and later balloon views), which depict "the town as seen obliquely from a more elevated point of view"; third are "linear ground plans, which showed the town from a theoretically vertical viewpoint"; and finally, there are "map views, in which the ground plan was enriched by delineations of detail in bird's-eye view. Map views had no vanishing point and therefore, strictly speaking, no perspective" (Hyde 11).

19. The inset maps of London and Westminster are those that appear in the *Speculum Britanniae. Civitas Londini* is shown as figure 4 in Hyde (42–43).

20. John Speed's inset map of Westminster in the corner of his map of Middlesex has the same orientation (with North-northeast at the top).

21. This feature makes it look as if the map exists *beneath* the landscape— just *waiting* to be revealed. Although he is discussing Norden's *Speculum Britanniae,* Lawrence Manley makes a point that seems relevant to this image of Westminster on the "long view": "The shape of Norden's London is no doubt a function of his segregation of Westminster for separate treatment elsewhere, yet this exclusion is itself a function of the connection he draws between the growth of London and the evolution of the constitutional mechanisms of the state" (1995, 157)

22. Interestingly, although Visscher never visited London, his "View" was to become one of the prototypes (along with Norden's) of many of the "views" that followed.

23. As in his fifteen-page history of the Tower in which he describes what each monarch added to the infrastructure; for example: "Edward IV. fortified the Tower of London, and enclosed with brick, as is aforesaid, a certain piece of ground, taken out of the Tower Hill, west from the Lion Tower, now called the bulwark. His officers also, in the 5th of his reign, set upon the said hill both scaffold and gallows, for the execution of offenders; whereupon the mayor, and his brethren complained to the king, and were answered that the same was not done in derogation of the city's liberties, and thereof caused proclamation to be made, &c., as shall be shown in Tower Street" (77, an episode important in demonstrating the strained relations between town and crown).

24. There was, as I have already indicated, a widespread practice of copying maps and views among those who had never been to London, but my point here is that initially *someone* had to take on the painstaking work of surveying and sketching for the original impression.

25. Or January 1558, using old-style dating.

26. This popularity seems to have been unexpected, since all but four pages had to be reset as well (Osborn, 17–22, "Bibliographical

Note" preceding *The Quenes Maiesties Passage*). It was also published in 1604, the year of James's royal entry (STC, 2nd ed. 7593). The account is usually attributed to Richard Mulcaster, although it was published anonymously: see Mullaney (154 n. 12), Wall (1993, 118), and Bergeron (1971, 15).

27. In citations for *The Quenes Maiesties Passage* I have used the page numbers from the Yale facsimile, edited by Osborn, as well as the signature notations.

28. Indeed, we cannot "look into all the roads and streets" as Braun and Hogenberg's directive for city *maps* requires.

29. Again, see Harley's discussion (1983, 31).

30. There is a long tradition of wronged women's poetic complaints, such as Ovid's *Heroides* and Chaucer's *Legend of Good Women*, although these are written by men.

31. Betty Travitsky (1980) and Wendy Wall (1991 and 1993, 299–310) have both discussed the poem in terms of the blazon tradition.

32. Again, Nancy Vickers's discussion of the blazon (outlined in chapter 3) is significant. See "'The Blazon of Sweet Beauty's Best': Shakespeare's Lucrece."

33. It would be odd if the prison were the place of sexual satisfaction, but the Counter prisons were apparently quite comfortable. Jonson, Marston, and Chapman's *Eastward Ho!* mentions "two Counters" (2.3.44) and C. G. Petter provides the following gloss: "the two debtors' prisons of London, under control of the sheriff, were at Wood Street and at Poultrey Street near St. Mildred's Church." According to Arthur Valentine Judges, "A man of means could live in comfort in either; indeed, for some people they served as a favorite retreat from the vengeance of enemies and even of the law. An apartment on the Master's Side of the prison was the best accommodation provided. The Knight's Ward was not so good, but comfortable as prison usage went. The Twopenny Ward and the Hole were no better than common jails—in some respects worse, for prisoners could count on no public assistance of any value in the provision of food, and a penniless man might actually starve to death in the Hole if he failed to secure relief or help from one of the citizens' legacies or the Christmas-treat funds provided for the very poor." (518). Another note regarding the Knights' Ward, the Hole, and the Twopenny Ward adds that the Counter had four types of accommodation, ranging in price from the Knights' Ward down to the Hole (Judges 423–87; see also Dekker and Webster's *Westward Ho!* 3.2.77–9). The "Hole" in the above note could also be the referent of the "certayne hole, and little ease within." "Counter," besides referring to the Counter Prison(s), could be meant to suggest the bawdy "counter"-blazons that were popular at this time as well.

34. Later, Thomas Tuke would condemn this same kind of nascent consumerism in his *A Treatise Against Painting and Tincturing of Men and Women:* "London, London hath her heart. The Exchange is the Temple of her Idols. In London she buys her head, her face, her fashion" (K2^v, qtd. in Howard 1994, 37).

    Jane Stevenson and Peter Davidson call the poem "a shopper's guide to Elizabethan London" (49).

35.   In Cheape of them, they store shal finde
                    and likewise in that streete:
              I Goldsmithes leaue, with Iuels such
                    as are for Ladies meete.
              And Plate to furnysh Cubbards with,
                    full braue there shall you finde:
              With Purle of Silver and of Golde,
                    to satisfye your minde.

                                                                   (99–106)

    [As if these *would* satisfy the mind.]

36.   In Cornwall, there I leaue you Beds,
                    and all that longs thereto.
                    .   .   .   .
              Artyllery at Temple Bar,
                    and Dagges at Tower hyll:
              Swords and Bucklers of the best
                    are nye the Fleete untyll.

                                                                 (125–138)

37.   Some Roysters styll, must bide in thee,
                    and such as cut it out:
              That with the guiltless quarel wyl,
                    to let their blood about.
              For them I cunning Surgions leaue,
                    some Playsters to apply.
              That Ruffians may not styll be hangde,
                    nor quiet persons dye.

                                                                 (147–154)

38. I think one could also argue that Whitney takes the body through stages of psychological development. Physical needs of food and clothing, sexual needs, and social or disciplinary needs (in the prisons). Freud's id, ego, superego as well as Abraham Maslow's hierarchy might be applied.

39. (British Library G 3685, transcribed by Richard Helgerson in *Forms of Nationhood,* 125). Here is another letter, also from Norden, but

addressed to Lord Burleigh, published at the front of Norden's *Preparative to his Speculum Britanniae:*

> To the Right Honourable, *Sir William Cecill Knight,* Baron of Burghleigh, Lord high Treasurer of England, of her Maiesties most Honourable priuie Counsell, and of the most Noble order of the Garter Knight.
>
> *Although (Right Honourable) I haue been forced, to strug-gle with want, the unpleasant companion of Industrious desires, and have long sustained foyle, inforced neglect of my purposed busines, and sorrow of my working spirit. It may yet now at the length please the high guide of Noble affections to moue your Ho-nour to effect what you haue begun: And as your hand hath happily led the way, your good worde may as easilye accomplish the worke of my newe reioycing. And the rather for that mine Indeuours in this generall businesse sprang from your Hon-ourable good liking. In regard wherof I am not in dispaire, but that my wythering hope shall be refreshed againe, with the dew of your powerfull helping hand, I haue under your patience and protection, upon some reasonable ground, exhibited this simple preparatiue unto the worlds view. And as I shall finde the same to answere your good opinion especially: so perforce will my heart and hand falter and fayle me, or fulfill what is hid in unseene desires.*
>
> Miseria mentem macerat.
At your Honours direction.
Iohn Norden.
>
> (A3ʳ⁻ᵛ, original spelling preserved, except long s's)

John Norden was able to complete only a few of the English coun-ties in his proposed *Speculum Britanniae.* William Smith, about whom less is known, may have suffered a similar fate; in any case, he began an atlas of county maps that was never completed (Barber II, 64–65).

40. This map is on the front cover of P. D. A. Harvey's *Maps in Tudor England.* It can be viewed online at:<http://www.tudorhistory.org/maps/cecilmap.jpg> and <http://www.tudorhistory.org/maps/cecilmapLG.jpg> (for high resolution).

41. Ironically, too, mapmakers often assign themselves a marginal posi-tion by drawing representations of themselves in the margins of their own maps.

42. The classic essay on the subject is J. W. Saunders, "The Stigma of Print: A Note on the Social Bases of Tudor Poetry" (1951). Saunders argues that since courtiers wrote their poetry to circulate among a coterie of their friends, professional poets who wanted to appear

genteel were faced with a quandary: "While the professional poet found printed books immeasurably advantageous to his social aspirations, he had to preserve as far as possible the illusion that he was genteel and well-bred, a suitable man for promotion into the activities of the Court and the noble houses and for the assumption of social responsibilities. To do this, he sought to imitate the manners and habits of the Court, and this in turn involved him in adopting, on the surface, their distaste for the publicity of print" (159).

43. Elizabeth Cary and others suggest that speaking in public might be worse than writing for the press; see Margaret W. Ferguson, "The Spectre of Resistance" (1991). For further discussion of the problems of women writers, see Wendy Wall, *The Imprint of Gender: Authorship and Publication in the English Renaissance* (1993) and Elaine V. Beilin, "Writing Public Poetry: Humanism and the Woman Writer" (1990).

44. *Sir Thomas More: Selected Letters,* ed. Elizabeth Frances Rogers (New Haven: Yale University Press, 1964: 151; quoted in Beilin, 1990, 250).

45. Whitney even "justifies" her own authorship according to this tradition earlier in her collection. In an epistolary poem in the *Nosegay,* Whitney writes to her sister:

> Had I a husband or a house,
> And all that [be]longes therto
> My selfe could frame about to rouse,
> As other women doo:
> But til some household cares me tye,
> My bookes and pen I wyll apply. (D2^r)

One reading of these lines would figure writing as an activity to fill a void for the unmarried woman; as such, it might well appease Whitney's married sister. Another reading would suggest that the "household cares" of the poem could be seen as an undesirable encumbrance or obstacle to writing.

46. The last two lines represent a peculiar moment of "giving" what must then be bought. Or perhaps Whitney is willing her equally needy friends the ability to buy.

47. Wendy Wall elaborates: "The woman's right to dispose of her own goods became a vexed issue during the 1550s. Local custom and common law had weakened the prohibition against married women making wills without the permission of their husbands," but apparently the state "began to crack down on the wife's right to make her own will at mid-century," according to Pearl Hogrefe, when Parliament enacted a statute in 1544 stating that "[w]ills or testaments made of manors, tenements, or other hereditaments, by any woman covert, by a person under twenty-one, by an idiot, or by any person

not sane, shall not be good under law" (Hogrefe 31, qtd. in Wall 1991, 45). Wall continues, "Although widows had more leeway in the matter, the property and goods of maids and wives belonged by law to the men who 'covered' them" (45). The categories of people deemed unfit to make a legitimate will, associating covert, or legally "covered," women with "idiot[s]," minors, and "person[s] not sane," reveals the magnitude of the problem of women and property.

48. The idea of the poem being a child, as suggested by the genre of the Mother's Legacy, provides an additional rationale for Whitney's beginning her delivery with "the head" of St. Paul's. Perhaps her crisscrossing the "matrix" of streets is also meant to hint at the child's formation in the womb. (The word "matrix," of course, has the Latin "mater" as its root and referred originally to the uterus and the child forming there.)

I have borrowed the phrase "paper landscape" from the title of a conference, "Paper Landscapes: Maps, Texts & the Construction of Space 1500–1700," organized by Andrew Gorden and Bernhard Klein at Queen Mary and Westfield College, University of London, July 1997, where I presented an earlier version of my work on Whitney.

49. And it could be that the "sheet" with which Whitney wants to be covered is a sheet of paper from that very book—thus making this pun yet another advertisement for her books.

Much has been written about Amelia Lanyer, another early modern writer, that helps us to understand the plight of the woman writer of the time. See Lanyer, Lewalski, McGrath, Mueller, and Schnell.

50. In *The Faerie Queene*, Spenser also dwells on the (female) body-as-place in the description of Alma's Castle (2.9). This description also tarries in the digestive parts, where the excremental functions are accorded thirteen lines (in stanzas 32–33). The fascination with this function is apparent when after a full-stanza description of pipes and conduits, Alma must lead the enthralled Guyon and Arthur away:

> Which goodly order, and great workmans skill
> Whenas those knights beheld, with rare delight,
> And gazing wonder their minds did fill;
> For neuer had they seene so straunge a sight.
> Thence backe againe faire *Alma* led them right,
>
> (2.9.33.1–5)

The "back gate" is also alluded to earlier in stanza 23's mention of "two gates" to the castle: "The one before, by which all in did pas" (the mouth), and "th'other" (the anus).

51. In fact, the relationship between the monarch and the city was always vexed, and all monarchs had to stop at Temple Bar and ask permission to enter the city. For an analysis of James I's less congenial coronation progress, see Goldberg (1983, 29–35).

52. The doctrine of the *femme sole* (to which category Elizabeth belongs, as queen, and to which Whitney *might* belong as unmarried woman without a father) could also apply here. In this light, the two women might be compared to the masterless men over whom much worry was expended at this time.

   Prostitutes, too, were ushered to the outskirts; I have already discussed how Whitney might fit this category (metaphorically) by virtue of her public speaking, writing, and taking money (figuratively, in the streets) for her books. Susan Frye also discusses the purse of money that Elizabeth receives from the city of London during her coronation progress in terms of prostitution (1993, 41–42).

53. One of Weimann's examples is Claudius on his throne at the *locus* and Hamlet soliloquizing on more generalized themes on the *platea*. Harry Berger makes use of Weimann's discussion to explicate scenes from *Henry IV, Part I* and *Part II*, and *The Two Gentlemen of Verona* (1998).

   In his more recent book chapter "Space (in)dividable: *locus* and *platea* revisited," Weimann adds examples from *Macbeth*, particularly the banquet and porter scenes, and *Timon of Athens*. He contends that

> the *locus* can be seen as a strategic approximation to the uses of perspectival form: it implicated the establishment of a topographically fixed locality. As in early modern cartography, the fixation itself, being in aid of knowledge and discovery, was instrumental in differentiating space and, thereby, separating the fixed place per abstraction from other localities. In the theatre, such fixation of different secular locations was conducive to an unprecedented specification and proliferation of symbolically encoded *loci*. (2000, 182)

54. "*Platea*" is also the classical Latin source of the word "place" (*OED*)

55. Indeed, in the larger Copperplate and Woodcut maps, we are privy to more of these specific actions. In Braun and Hogenberg's rendition, besides the four large foreground figures, the activity that is most visible is that which takes place on and near the river.

56. There is at times some blurring of the distinction between *locus* and *platea*. Here the *locus* is the everyday stomping grounds of the audience who are adjacent to the *platea* and to its "comic" depiction of London's citizens. A misunderstanding of those boundaries is at the heart of Francis Beaumont's *The Knight of the Burning Pestle* (ca. 1610), where a citizen from the audience and his wife want their man

Rafe to play the hero. The same misunderstanding is often central to the antitheatrical tracts, as we shall see.

Likewise, the demarcation is blurred in the epilogue to *Eastward Ho!* when Quicksilver, emerging from the Counter prison, imagines the theater and its audience as the streets of London and the fronts of houses:

> Stay, sir, I perceive the multitude are gathered together to view our coming out at the Counter. See, if the streets and the fronts of the houses be not stuck with people, and the windows filled with ladies, as on the solemn day of the Pageant! (Epilogue 1–5)

57. In *The City Staged: Jacobean Comedy, 1603–1613,* Leinwand concerns himself especially with the triangulation of merchants, gallants, and women (as wives, widows, whores, and maids).

58. In "An Exact and perfect CATALOGUE of all the PLAIES that were ever printed; together, with all the Authors names . . ." (appended to the 1656 quarto of *The Old Law*), the publisher assigns *Michaelmas Term* to Middleton. Although in a "source that is far from reliable," the attribution has not been seriously questioned, according to Richard Levin, because of its similarity to other city comedies by Middleton performed by the Children of Paul's (Levin x).

59. This dressing up may again remind us of the "dressing up" of the city for Queen Elizabeth.

60. The pun on "foot" and the French "foutre," always available for early modern texts, reinforces the link between sexual impropriety and the shady acquisition of land that is one hallmark of city comedy.

61. Shortyard's punning name belies his own impotence or "short-comings," since "yard" meant penis in the bawdy parlance of the day.

62. The names of Quomodo's assistants, Shortyard and Falselight, allude to the practices of using inaccurate measuring rods and dim lighting to cheat customers. "Yard" here refers to both the cloth Quomodo might be measuring and to the slang term for penis referred to above, with implications of sexual congress as well as masturbation in "measuring with his yard."

63. See Joseph Cady, "'Masculine Love,' Renaissance Writing, and the 'New Invention' of Homosexuality," for a discussion of the anachronistic use of the terms "homosexuality" and "homoeroticism."

64. Although this setting is implied by a title that refers to a particular court session, most scenes in the play are actually set elsewhere in the city.

65. Although it is never mentioned in the play, St. Bartholomew's Day, the time frame of the play, is also one of the traditional days for "beating the bounds" of a property or a parish. Here those boundaries are not only the physical boundaries of the city, but the boundaries of decorum, religion, and representation.

66. Peter Stallybrass and Allon White describe her as "belly, womb, gaping mouth, udder, the source and object of praise and abuse. Above all, like the giant hog displayed at the fair, she is *excessive*" (64).

67. Here, Jonson may be following the antitheatricalists such as Stephen Gosson, when he warns: "Thought is free; you can forbidd no man, that vieweth you, to noute you and that noateth you, to judge you, for entring to places of suspition" (*Schoole of Abuse*, F2^r, qtd. in Howard 1991, 71).

68. All of the holes that motivate the action of the play (with the exception of the stocks) are orifices of women's bodies; the grotesque body, it seems, has taken over. Win's womb causes the craving for pork, her need to urinate brings Win and her group to Ursula's box where she is tricked out like a prostitute. Gail Kern Paster offers a vigorous discussion of the problem of women as leaky vessels in this and other plays in *The Body Embarassed* (23–63). See also Andrew McRae's excellent essay on Ben Jonson's "On the Famous Voyage" for an account of a navigation up the polluted Fleet Ditch.

69. "Vapour" here means "a fancy or fantastic idea; a foolish brag or boast" (*OED*, definition 4, where two other passages from *Bartholomew Fair* are cited as examples).

70. According to Phillip Stubbes, plays were "ordeined by the Devill, and consecrat to heathen Gods, to draw us from Christianitie to ydolatrie, and gentilisme" (L6^v).

71. The pageants also cover over some internal conflicts that may exist within the guilds. The upper body of liveried traders and the larger group of craftsman experienced much tension. See Lobanov-Rostovsky for a further discussion of these problems (889–890).

72. Error has quite a lengthy speech, in fact, in which she greets the new mayor and suggests ironically that Virtue "hath so long impoverish'd this fair city" (Bullen VII: 242), and warns about "a poor, thin, threadbare thing call'd Truth"; she bawdily suggests that Truth

> H'as but one foolish way, straight on, right forward,
> And yet she makes a toil on't, and goes on
> With care and fear, forsooth, when I can run
> Over a hundred with delight and pleasure,
> Back-ways and by-ways, and fetch in my treasure
> After the wishes of my heart, by shifts,
> Deceits, and slights: and I'll give thee those gifts;
> I'll show thee all my corners yet untold,
> The very nooks where beldams hide their gold,
> In hollow walls and chimneys, where the sun
> Never yet shone, nor Truth came ever near:
> This of thy life I'll make the golden year;
> Follow me then. (Bullen, Vol. VII, 243)

Sergei Lobanov-Rostovsky gives an extended reading of this
pageant in which he suggests that some of the anxiety over the
change of power is played out here (885–888). Theodore Leinwand
suggests, however, that "what Error has to offer, the Lord Mayor already has" (1982, 146).

73. Bergeron's source is the Wells Town Clerk's Office, *Wells Acts of the
Corporation 1553–1623*, fol. 374.

EPILOGUE

1. The image measures 241.3 X 152.4 cm, or about 8' in height X 5' in
width.

2. This type of imagery is very much with us today; Sharon Olds's poem
"Topography" compares two lovers to "maps laid / face to face," detailing which parts of the lovers' bodies touch: "your Fire Island
against my Sonoma," and so on. I am reminded especially of Michael
Franks's song "Popsicle Toes," which includes the lines,

> You've got the nicest North America this sailor ever saw,
> I'd like to feel your warm Brazil and touch your Panama.
> But your Tierra del Fuegos are nearly always froze,
> We've got to see-saw until we unthaw those popsicle toes.
> —Excerpt from "Popsicle Toes"
> copyright © 1975 by Michael Franks,
> reprinted by permission of Mississippi Mud Music

3. Angela Carter is fond of describing St. Paul's as "the single Amazon
breast" of London (*Nights at the Circus* 291; *Wise Children* 3). Of
course, in Elizabeth's day, St. Paul's was not domed, but perhaps
Carter means to make her reference yet another remembrance of that
"Amazon queen" in her native topography.

4. Another deck of cards from 1676, known as Robert Morden's Playing Cards, offers better maps, but a deceiving scale, since all of the
counties are roughly the same size. They offer information on size,
but not much else; gone is the doggerel verse of the earlier cards
along with its information about the counties.

BIBLIOGRAPHY

A discourse concerning the drayning of fennes and surrounded grounds. [facs]. Amsterdam: Theatrum Orbis Terrarum, 1629.

Adams, Robert. *Expeditionis Hispanorum in Angliam vera descriptio,* 1588.

Adams, William Howard. *Nature Perfected: Gardens Through History.* New York: Abbeville Press, 1991.

Agas, Radolph. *A Preparative to Platting of Landes and Tenements for Surveigh. Shewing the Diversitie of sundrie instruments applyed therunto. Patched up as plainly together, as boldly offered to the curteous view and regard of all worthie Gentlemen, lovers of skill. And published in stead of his flying papers, which cannot abide the pasting to poasts.* London: Thomas Scarlet, 1596.

Agas, Radolph. *To all persons whom these presents may concerne, of what estate and degree soeuer by Radulph Agas . . .* London, 1596.

Alpers, Svetlana. *The Art of Describing: Dutch Art in the Seventeenth Century.* Chicago: University of Chicago Press, 1983.

Alpers, Svetlana. "The Mapping Impulse in Dutch Art." In *Art and Cartography: Six Historical Essays.* Ed. David Woodward. Chicago and London: University of Chicago Press, 1987, pp. 51–96.

Althusser, Louis. "Ideology and Ideological State Apparatuses." In *Critical Theory Since 1965.* Ed. Hazard Adams and Leroy Searle. Tallahassee, Florida: Florida State University Press, 1970, pp. 239–250.

Anderson, Benedict. *Imagined Communities: Reflections on the Origin and Spread of Nationalism.* Revised ed. London and New York: Verso, 1991.

Andrews, J. H. "Appendix: the beginnings of the surveying profession in Ireland—abstract." In *English Map-Making 1500–1650.* Ed. Sarah Tyacke. London: The British Library, 1983, pp. 20–21.

Archer, Ian. "The Nostalgia of John Stow." In *The Theatrical City: Culture, Theatre and Politics in London, 1576–1649.* Ed. David L. Smith, Richard Strier, and David Bevington. Cambridge and New York: Cambridge University Press, 1995, pp. 17–34.

Bacon, Sir Francis. *The Advancement of Learning and The New Atlantis.* London: Oxford University Press, 1966.

Bagrow, Leo. *History of Cartography.* 2nd ed. Chicago: Precedent Publishers, (1963) 1985.

Bakhtin, Mikhail. *Rabelais and His World.* Trans. Helene Iswolsky: Indiana University Press, 1984.

Baptiste, Victor N. *Bartolome de Las Casas and Thomas More's* Utopia. Culver City, Calif.: Labyrinthos, 1990.

Barber, Peter. "The Copperplate Map in Context." In *Tudor London: A Map and a View*. Ed. Ann Saunders and John Schofield. London: London Topographical Society, Publication no. 159, in association with the Museum of London, 2001, pp. 16–32.

Barber, Peter. "England I: Pageantry, Defense, and Government: Maps at Court to 1550." In *Monarchs, Ministers, and Maps: The Emergence of Cartography as a Tool of Government in Early Modern Europe*. Ed. David Buisseret. Chicago: University of Chicago Press, 1992, pp. 26–56.

Barber, Peter. "England II: Monarchs, Ministers, and Maps, 1550–1625." In *Monarchs, Ministers, and Maps: The Emergence of Cartography as a Tool of Government in Early Modern Europe*. Ed. David Buisseret. Chicago: University of Chicago Press, 1992, pp. 57–98.

Barish, Jonas A., ed. *Ben Jonson: A Collection of Critical Essays*. Englewood Cliffs, N.J.: Prentice-Hall, 1963.

Bartolovich, Crystal. "Spatial Stories: *The Surveyor* and the Politics of Transition." In *Place and Displacement in the Renaissance*. Ed. Alvin Vos. Binghamton, N.Y.: Medieval and Renaissance Studies, Vol. 132, 1995, pp. 255–283.

Baudrillard, Jean. *Simulations*. Trans. Paul Foss, Paul Patton, and Philip Beitchman. New York: Semiotext(e), Inc., 1983.

Beaumont, Francis. *The Knight of the Burning Pestle. Drama of the English Renaissance II: The Stuart Period*. Ed. Russell A. Fraser and Norman Rabkin. London and New York: Macmillan, 1976.

Beilin, Elaine V. *Redeeming Eve: Women Writers of the English Renaissance*. Princeton, N.J.: Princeton University Press, 1987.

Beilin, Elaine V. "Writing Public Poetry: Humanism and the Woman Writer." *Modern Language Quarterly* 51:2 (1990): 249–71.

Bellamy, Elizabeth. "The Vocative and the Vocational: The Unreadability of Elizabeth in *The Faerie Queene*." *ELH* 54:1 (1987): 1–30.

Bentham, Jeremy. *The Panopticon Writings*. Ed. Miran Bozovic. London and New York: Verso, 1787.

Berger, Harry, Jr. "The Prince's Dog: Falstaff and the Perils of Speech-Prefixity." *Shakespeare Quarterly* 49:1 (1998): 40–73.

Berger, Harry, Jr. *Revisionary Play: Studies in the Spenserian Dynamics*. Berkeley: University of California Press, 1988.

Bergeron, David. *English Civic Pageantry: 1558–1642*. London: Edward Arnold, 1971.

Bergeron, David. *Practicing Renaissance Scholarship: Plays and Pageants, Patrons and Politics*. Pittsburgh: Duquesne University Press, 2000.

Berry, Philippa. *Of Chastity and Power: Elizabethan Literature and the Unmarried Queen*. London and New York: Routledge, 1994.

Blake, Erin. "Where Be '*Here be Dragons*'? Ubi sunt '*Hic sunt dracones*'?" *MapHist Home Page, University of Utrecht* (1999): <http://www.maphist.nl/extra/herebedragons.html>.

Boccaccio, Giovanni. "The Second Day, The Ninth Novel." In *The De-cameron*. Trans. unknown. London: Isaac Jaggard, 1620.

Boesky, Amy. *Founding Fictions: Utopias in Early Modern England*. Athens and London: University of Georgia Press, 1996.

Booth, Stephen. *Shakespeare's Sonnets*. New Haven: Yale University Press, 1977.

Booty, John, ed. *The Book of Common Prayer*, 1559.

Borges, Jorge Luis. *Dreamtigers*. Trans. Mildred Boyer and Harold Morland. Austin: University of Texas Press, 1964.

Borges, Jorge Luis. *Historia universal de la infamia*. Buenos Aires: Emecé Editores, 1954.

Borges, Jorge Luis. *A Universal History of Infamy*. Trans. Norman Thomas di Giovanni. New York: Dutton, 1972.

Bourdieu, Pierre. *Outline of a Theory of Practice*. Cambridge: Cambridge University Press, 1977.

Bowes, W. *Playing Cards, Memory Cards*. London, ca. 1605.

Braun, Georg, and Frans Hogenberg. *Civitates Orbis Terrarum*, 1572.

Breight, Curtis C. *Surveillance, Militarism and Drama in the Elizabethan Era*. New York: St. Martin's Press, 1996.

Breight, Curtis C. "'Treason doth never prosper': *The Tempest* and the Dis-course of Treason." *Shakespeare Quarterly* 41:1 (1990): 1–28.

Brooke, Ralph. *A Discoverie of certaine errours published in the much-com-mended Britannia*, 1594.

Bry, Theodore de. *The Discovery of the New World: Engravings by Th. de Bry*. Amsterdam: Van Hoeve, 1979.

Bucher, Bernadette. "De Bry's *Grand Voyages* (1590–1634): The first grand-scale European reportage on America." In *America, Bride of the Sun*. Ed. Royal Museum of Fine Arts Antwerp. Brussels: Imschoot Books, 1992, pp. 129–140.

Bucher, Bernadette. *Icon and Conquest: A Structural Analysis of the Illustra-tions of de Bry's Great Voyages*. Trans. Basia Miller Gulati. Chicago: University of Chicago Press, 1981.

Bucher, Bernadette, Rolena Adorno, and Mercedes López-Baralt. *La Icono-grafía Política del Nuevo Mundo*. Río Piedras, Puerto Rico: Universidad de Puerto Rico, 1990.

Buisseret, David, ed. *Monarchs, Ministers, and Maps: The Emergence of Car-tography as a Tool of Government in Early Modern Europe*. Chicago and London: University of Chicago Press, 1992.

Bullen, A. H., ed. *The Works of Thomas Middleton*. 8 Vols. London: John C. Nimmo, 1895.

Buncombe, Marie H. "Faire Florimell as Faire Game: The Virtuous, Un-married Woman in *The Faerie Queene* and *The Courtier*." *CLA Journal* 28:2 (1984): 164–175.

Butler, Judith. *Gender Trouble: Feminism and the Subversion of Identity*. New York: Routledge, 1990.

Cady, Joseph. "'Masculine Love,' Renaissance Writing, and the 'New In-vention' of Homosexuality." In *Homosexuality in Renaissance and*

Enlightenment England. Ed. Claude Summers. New York and London: Haworth, 1992, pp. 9–40.

Camden, William. *Britànnia,* 1594.

Camões, Luis Vaz de. *The Lusiads.* Trans. William C. Atkinson. London: Penguin, 1572.

Campbell, Mary. *The Witness and the Other World: Exotic European Travel Writing 400–1600.* Ithaca: Cornell University Press, 1988.

Campion, Thomas. "Observations in the Art of English Poesie." In *Elizabethan Critical Essays.* Ed. G. Gregory Smith. 1904 ed. London: Oxford University Press, 1602. Vol. II, pp. 327–355.

Carroll, Clare. "Representations of Women in Some Early Modern English Tracts on the Colonization of Ireland." *Albion* 25:3 (1993): 379–393.

Carruthers, Mary. *The Book of Memory: A Study of Memory in Medieval Culture.* Cambridge: Cambridge University Press, 1990.

Carter, Angela. *Nights at the Circus.* New York and London: Penguin, 1984.

Carter, Angela. *Wise Children.* New York and London: Penguin, 1991.

Casey, Edward S. *The Fate of Place: A Philosophical History.* Berkeley and Los Angeles: University of California Press, 1997.

Castiglione, Baldasare. *The Boke of the Courtier from the Italian of Count Baldasare Castiglione Done into English by Sir Thomas Hoby.* London, 1561.

Castiglione, Baldesar. *The Book of the Courtier.* Trans. George Bull. London: Penguin, 1967.

Cavanagh, Sheila T. *Wanton Eyes and Chaste Desires: Female Sexuality in* The Faerie Queene. Bloomington: Indiana University Press, 1994.

Caxton, William. *Caxton, The Description of Britain: A Modern Rendering by Marie Collins.* New York: Weidenfeld & Nicolson, 1988.

Certeau, Michel de. *The Practice of Everyday Life.* Trans. Steven Rendell. Berkeley: University of California Press, 1984.

Certeau, Michel de. *The Writing of History.* Trans. Tom Conley. New York: Columbia University Press, 1988.

Chambers, E. K. *The Elizabethan Stage.* 4 Vols. Oxford: Clarendon, 1923.

Chandler, John, ed. *John Leland's Itinerary: Travels in Tudor England.* Dover, N. H.: Alan Sutton, 1993.

Cheney, Patrick. "'And Doubted Her to Deeme an Earthly Wight': Male Neoplatonic 'Magic' and the Problem of Female Identity in Spenser's Allegory of the Two Florimells." *Studies in Philology* 86: (1989): 310–340.

Cheney, Patrick. *Marlowe's Counterfeit Profession: Ovid, Spenser, Counternationhood.* Toronto, Buffalo, and London: University of Toronto Press, 1997.

Christian, Roy. *Old English Customs.* New York: Hastings House, 1966.

Cicero, Marcus Tullius. *De oratore.* Trans. E. W. Sutton. Cambridge: Harvard University Press, 1942.

Clark, Glenn. "The 'Strange' Geographies of *Cymbeline.*" In *Playing the Globe: Genre and Geography in English Renaissance Drama.* Ed. John Gillies and Virginia Mason Vaughan. Madison and Teaneck, N.J. and

London: Fairleigh Dickinson University Press and Associated University Presses, 1998, pp. 230–259.

Cliffe, J. T. *The World of the Country House in Seventeenth-Century England.* New Haven: Yale University Press, 1999.

Clifford, D. J. H., ed. *The Diaries of Lady Anne Clifford.* 1991 ed. Wolfboro Falls, N. H.: Alan Sutton, 1990.

Coiro, Ann Baynes. "Writing in Service: Sexual Politics and Class Position in the Poetry of Aemilia Lanyer and Ben Jonson." *Criticism* 35: (1993): 357–376.

Columbus, Christopher. *Select Documents Illustrating the Four Voyages of Columbus.* Trans. Cecil Jane. London: Hakluyt Society, 1930.

Conley, Tom. *The Self-Made Map: Cartographic Writing in Early Modern France.* Minneapolis: University of Minnesota Press, 1996.

Cook, Olive. *The English Country House: An Art and a Way of Life.* London: Thames and Hudson, 1974.

Cooper, Quentin and Paul Sullivan. *Maypoles, Martyrs & Mayhem: 366 Days of British Customs, Myths, and Eccentricities.* London: Bloomsbury Publishing, 1994.

Crewe, Jonathan. "The Garden State: Marvell's Poetics of Enclosure." In *Enclosure Acts: Sexuality, Property, and Culture in Early Modern England.* Ed. Richard Burt and John Michael Archer. Ithaca and London: Cornell University Press, 1994, pp. 270–289.

Cubeta, Paul M. "A Jonsonian Ideal: 'To Penshurst.'" *Philological Quarterly* XLII (Jan):1 (1963): 14–24.

Cuningham, William. *The Cosmographical Glass.* The English Experience: Its Record in Early Printed Books Published in Facsimile, Number 44. 1559 [facsimile] ed. Amsterdam and New York: Theatrum Orbis Terrarum, Ltd. and De Capo Press, 1968.

Daniel, Samuel. "A Defence of Ryme, Against a Pamphlet entituled: Observations in the Art of English Poesie: Wherein is demonstatively prooved, that Ryme is the fittest harmonie of words that comports with our language." In *Complete Works in Verse and Prose.* 5 Vols. Ed. Alexander B. Grosart. New York: Russell and Russell, 1963., pp. 35–67.

Davies, Sir John. *Hymnes of Astraea. The Complete Poems of Sir John Davies.* Ed. Alexander B. Grosart. 1869 ed. Ann Arbor: University Microfilms, 1969.

Dekker, Thomas. *The Dramatic Works of Thomas Dekker.* Ed. Fredson Bowers. Cambridge: Cambridge University Press, 1966.

Delano-Smith, Catherine, and Roger J. P. Kain. *English Maps: A History.* Toronto and Buffalo: University of Toronto Press, 1999.

Devi, Mahasweta. "Douloti the Bountiful." In *Imaginary Maps.* Ed. Gayatri Chakravorty Spivak. Trans. Gayatri Chakravorty Spivak. New York: Routledge, 1994, pp. 19–93.

Donne, John. *John Donne.* Ed. John Carey. Oxford and New York: Oxford University Press, 1990.

Donovan, Josephine. "Toward a Women's Poetics." In *Feminist Issues in Literary Scholarship*. Ed. Shari Benstock. Bloomington: Indiana University Press, 1987, pp. 98–109.

Drayton, Michael. *Poly-Olbion*, Facsimile of 1613 edition. Works of Michael Drayton. ed. J. William Hebel. Vol 4. (5 Vols.). 1933.

Drayton, Michael. *Works, Volume IV: Poly-Olbion*. Ed. J. William Hebel. Oxford: Shakespeare Head Press, 1613.

Dymock, Cressy. *A Discoverie For Division or Setting out of Land, as to the best Form. Published by Samuel Hartlib Esquire, for Direction and more Advantage and Profit of the Adventurers and Planters in the FENS and other Waste and undisposed Places in England and IRELAND. Whereunto are added some other Choice Secrets or Experiments of Husbandry. With a Philosphical Quere concerning the Cause of Fruitfulness. AND An Essay to Shew How all Lands may be improved in a New Way to become the ground of the increase of Trading and Revenue to this Common-wealth*. London: Richard Wodenotbe, 1653.

Eco, Umberto. "On the Impossibility of Drawing a Map of the Empire on a Scale of 1 to 1." In *How to Travel with a Salmon & Other Essays*. Trans. William Weaver. New York: Harcourt Brace, 1994, pp. 95–106.

Edgerton, Samuel Y. Jr. "From Mental Matrix to *Mappamundi* to Christian Empire: The Heritage of Ptolemaic Cartography in the Renaissance." In *Art and Cartography: Six Historical Essays*. Ed. David Woodward. Chicago and London: University of Chicago Press, 1987, pp. 10–50.

Elliot, James. *The City in Maps: Urban Maps to 1900*. London: The British Library, 1987.

Elyot, Sir Thomas. *The Boke Named the Governour*. Ed. S. E. Lehmberg. London, 1962.

Evans, G. Blakemore, ed. *The Riverside Shakespeare*. 2nd ed. Boston and New York: Houghton Mifflin Company, 1997.

Falco, Raphael. *Conceived Presences: Literary Genealogy in Renaissance England*. Amherst: University of Massachusetts Press, 1994.

Farmer, Norman K., Jr. "The World's New Body: Spenser's *Faerie Queene* Book II, St Paul's Epistles and Reformation England." In *Renaissance Culture in Context*. Ed. Jean R. Brink and William F. Gentrup. Aldershot, England: Scolar Press, 1993, pp. 75–85.

Ferguson, Margaret W. "The Spectre of Resistance: *The Tragedy of Mariam* (1613)." In *Staging the Renaissance: Reinterpretations of Elizabethan and Jacobean Drama*. Ed. David Scott Kastan and Peter Stallybrass. New York and London: Routledge, 1991, pp. 235–250.

Fischlin, Daniel. "Political Allegory, Absolutist Ideology, and the 'Rainbow Portrait' of Queen Elizabeth I." *Renaissance Quarterly* 50:1 (1997): 175–206.

Fitzherbert, John. *The Boke of Surveying and Improvements*. Facsimile of 1523 ed. The English Experience. Amsterdam and Norwood, N.J.: Theatrum Orbis Terrarum, Ltd. and Walter J. Johnson, Inc., 1974.

Foucault, Michel. *Discipline and Punish: the Birth of the Prison*. Trans. Alan Sheridan. New York: Vintage, 1979.

Foucault, Michel. "Of Other Spaces." *Diacritics* Spring: (1986): 22–27.

Foucault, Michel. "Questions on Geography." In *Power/Knowledge: Selected Interviews and Other Writings 1972–1977*. Ed. Colin Gordon. Trans. Colin Gordon, Leo Marshall, John Mepham, and Kate Soper. New York: Pantheon, 1980, pp. 63–77.

Fowler, Alastair. *The Country House Poem: A Cabinet of Seventeenth-Century Estate Poems and Related Items*. Edinburgh: Edinburgh University Press, 1994.

Fowler, Alastair. "Georgic and Pastoral: Laws of Genre in the Seventeenth Century." In *Culture and Cultivation in Early Modern England: Writing and the Land*. Ed. Michael Leslie and Timothy Raylor. Leicester and London: Leicester University Press, 1992, pp. 81–88.

Fowler, Alastair. *Spenser and the Numbers of Time*. London: Routledge & Kegan Paul, 1964.

Foxe, John. *Actes and Monuments of matters most speciall and memorable, happening in the Church, with an universall history of the same . . .* London: Peter Short, 1596.

Foxe, John. *Acts and Monuments*. 8 Vols. New York: AMS Press, 1965.

Fraser, Russell A., and Norman Rabkin, eds. *Drama of the English Renaissance, Volume I: The Tudor Period*. New York: Macmillan, 1976.

Fraser, Russell A., and Norman Rabkin, eds. *Drama of the English Renaissance, Volume II: The Stuart Period*. New York: Macmillan, 1976.

Friedman, Alice T. "Wife in the English Country House: Gender and the Meaning of Style in Early Modern England." In *Women and Art in Early Modern Europe: Patrons, Collectors, and Connoisseurs*. Ed. Cynthia Lawrence. University Park, Pennsylvania: Pennsylvania State University Press, 1997, pp. 111–125.

Frye, Susan. *Elizabeth I: The Competition for Representation*. New York and London: Oxford University Press, 1993.

Frye, Susan. "Of Chastity and Violence: Elizabeth I and Edmund Spenser in the House of Busirane." *Signs: Journal of Women in Culture and Society* 20:1 (1994): 49–78.

Gascoigne, George. *A briefe rehearsall, or rather a true copie of as much as was presented before Her Majesties at Kenelworth, during her last aboade there, as followeth. The whole woorkes of George Gascoinge, Esquyre*. London: Abell Jeffes, 1587.

Gellner, Ernest. *Nations and Nationalism*. Ithaca: Cornell University Press, 1983.

Giamatti, A. Bartlett. *Play of Double Senses: Spenser's* Faerie Queene. New York: Norton, 1975.

Gibbons, Brian. *Jacobean City Comedy: A Study of Satiric Plays by Jonson, Marston, and Middleton*. Cambridge: Harvard University Press, 1968.

Gillies, John. *Shakespeare and the Geography of Difference.* Cambridge: Cambridge University Press, 1994.

Gilson, J. P., ed. *A Summary of the Records and a True Memorial of the Life of Me the Lady Anne Clifford,* n.d.

Girard, René. *Deceit, Desire, and the Novel: Self and Other in Literary Structure.* Trans. Yvonne Freccero. Baltimore: Johns Hopkins University Press, 1972.

Glaser, Lynn. *America on Paper: The First Hundred Years.* Philadelphia: Associated Antiquaries, 1989.

Glover, Moses. *Istleworth Hunderd, Being The Mannor of SION And one of The Seauen Hunderds in Comita MIDDLESEX: TOTALLY Described: With All the TOWNES & Villages Highewayes, Brookes, Commones & perticuler Incloshures, With theyr true Formes and Contentes of Akers Roodes and perches; And Whatsoever else Remarkable, Excectly Calculated, wth Methematecall Instruments, By angles and Figures, according to the Rulles of the most Famous Science of Geometry; Being one of the Lordshipps, and parte of the Reuenues of that potente peere and truely Honoured ALGERNOUN PERCY; EARLE of NORTHUMBERLAND. My Noble Lord & Master Hielde of our SOVRAYNE Lorde the KINGE in fee farme By Letter patienes Bareing date.* Facsimile of 1635 ed. Ditchling, Sussex, England: Pitkin Pictorials Ltd. [n.d.].

Goldberg, Jonathan. *Desiring Women Writing: English Renaissance Examples.* Stanford: Stanford University Press, 1997.

Goldberg, Jonathan. *Endlesse Worke: Spenser and the Structures of Discourse.* Baltimore and London: Johns Hopkins University Press, 1981.

Goldberg, Jonathan. *James I and the Politics of Literature: Jonson, Shakespeare, Donne, and Their Contemporaries.* Baltimore and London: Johns Hopkins University Press, 1983.

Goldberg, Jonathan. "'They Are All Sodomites': The New World." In *Sodometries: Renaissance Texts, Modern Sexualities.* Stanford: Stanford University Press, 1992, pp. 179–249.

Goodwin, Barbara, and Keith Taylor. *The Politics of Utopia: A Study in Theory and Practice.* New York: St. Martin's, 1982.

Gorton, Lisa. "John Donne's Use of Space." *Early Modern Literary Studies* 4.2 / Special Issue 3 (1998): 9.1–27 <URL: http://purl.oclc.org/emls/04–2/gortjohn.htm>.

Gosson, Stephen. *The Schoole of Abuse, Conteining a Pleasaunt Invective against Poets, Pipers, Plaiers, Jesters, and such like Caterpillers of a Commonwealth.* London, 1579.

Gould, Peter and Rodney White. *Mental Maps.* 2nd ed. Boston: Allen & Unwin, 1986.

Greenblatt, Stephen. "Learning to Curse: Aspects of Linguistic Colonialism in the Sixteenth Century." In *Learning to Curse: Essays in Early Modern Culture.* 1992 ed. New York: Routledge, 1990, pp. 16–39.

Greenblatt, Stephen. *Marvelous Possessions: The Wonder of the New World.* Chicago: University of Chicago Press, 1991.

Greenblatt, Stephen, ed. *New World Encounters*. Berkeley: University of California Press, 1993.

Greenblatt, Stephen and Giles Gunn, ed. *Redrawing the Boundaries: The Transformation of English and American Literary Studies*. New York: Modern Language Association, 1992.

Greene, Thomas. *The Light in Troy: Imitation and Discovery in Renaissance Poetry*. New Haven and London: Yale University Press, 1982.

Greenlaw, Edwin, Charles Grosvenor Osgood, and Frederick Morgan Padelford, eds. *The Works of Edmund Spenser, A Variorum Edition*. 11 Vols. Baltimore: Johns Hopkins, 1966.

Greer, Germaine, et al., ed. *Kissing the Rod: An Anthology of Seventeenth-Century Women's Verse*. New York: The Noonday Press, 1988.

Guillory, John. *Poetic Authority: Spenser, Milton, and Literary History*. New York: Columbia University Press, 1983.

Hadfield, Andrew. *Literature, Politics and National Identity: Reformation to Renaissance*. Cambridge: Cambridge University Press, 1994.

Hakluyt, Richard. "Discourse of Western Planting." In *The Original Writings & correspondence of the Two Richard Hakluyts*. Ed. E. G. R. Taylor. 1935 ed. London: The Hakluyt Society, 1584. Vol. LXXVII.

Hakluyt, Richard. *The Principal Navigations, Voyages, Traffiques and Discoveries of the English Nation*. 12 Vols. 1903–1905 ed. Glasgow: J. MacLehose & Sons, 1589.

Halpern, Richard. *The Poetics of Primitive Accumulation: English Renaissance Culture and the Genealogy of Capital*. Ithaca: Cornell University Press, 1991.

Hamilton, A. C., ed. *The Faerie Queene*. London and New York: Longman, 1977.

Hamilton, A. C. *The Structure of Allegory in* The Faerie Queene. Oxford: Clarendon Press, 1961.

Harley, J. B. "Deconstructing the Map." *Cartographia* 26:2 (1989): 1–20.

Harley, J. B. "The Map and the Development of the History of Cartography." In *The History of Cartography: Volume 1, Cartography in Prehistoric, Ancient, and Medieval Europe and the Mediterranean*. Ed. J. B. Harley and David Woodward. Chicago and London: University of London Press, 1987, pp. 1–42.

Harley, J. B. "Maps, Knowledge, and Power." In *The Iconography of Landscape*. Ed. Denis Cosgrove and Stephen Daniels. Cambridge: Cambridge University Press, 1988, pp. 277–312.

Harley, J. B. "Meaning and Ambiguity in Tudor Cartography." In *English Map-Making 1500–1650*. Ed. Sarah Tyack. London: The British Library, 1983, pp. 22–45.

Harley, J. B. "Silences and Secrecy: The Hidden Agenda of Cartography in Early Modern Europe." *Imago Mundi* 40: (1988): 57–76.

Harley, J. B. and David Woodward. *The History of Cartography*. 6 Vols. Chicago and London: University of Chicago Press, 1987.

Harriot, Thomas. *A Briefe and True Report of the New Found Land of Virginia.* Facsimile of 1590 ed. New York: Dover Publications, 1972.

Harris, John. *The Design of the English Country House: 1620–1920.* London: Trefoil Books, 1985.

Harris, Jonathon Gil. "This Is Not a Pipe: Water Supply, Incontinent Source, and the Leaky Body Politic." In *Enclosure Acts: Sexuality, Property, and Culture in Early Modern England.* Ed. Richard Burt and John Michael Archer. Ithaca: Cornell University Press, 1994, pp. 203–228.

Harrison, William. *The Description of England: The Classic Contemporary Account of Tudor Social Life.* Ed. Georges Edelen. 1994 ed. Washington, D.C. and New York: Folger Shakespeare Library and Dover, 1964.

Hartog, François. *The Mirror of Herodotus: The Representation of the Other in the Writing of History.* Trans. Janet Lloyd. Berkeley: University of California Press, 1988.

Harvey, David. *The Condition of Postmodernity: An Enquiry into the Origins of Cultural Change.* Cambridge, Mass. and Oxford: Blackwell, 1990.

Harvey, P. D. A. "Estate Surveyors and the Spread of the Scale-map in England 1550–80." *Landscape History: Journal of the Society for Landscape Studies* 15: (1993): 37–49.

Harvey, P. D. A. *Mappa Mundi: The Hereford World Map.* Toronto and Buffalo: University of Toronto Press, 1996.

Harvey, P. D. A. *Maps in Tudor England.* Chicago: University of Chicago Press, 1993.

Harvey, P. D. A. *Medieval Maps.* London: The British Library, 1991.

Heal, Felicity and Clive Holmes. *The Gentry in England and Wales, 1500–1700.* Stanford: Stanford University Press, 1994.

Helgerson, Richard. *Forms of Nationhood: The Elizabethan Writing of England.* Chicago: University of Chicago Press, 1992.

Helgerson, Richard. "Introduction to Special Issue on Literature and Geography." *Early Modern Literary Studies* 4.2 / Special Issue 3 (1998): 1.1–14 <http://www.shu.ac.uk/emls/04–2/intro.htm>.

Helgerson, Richard. *Self-Crowned Laureates: Spenser, Jonson, Milton, and the Literary System.* Berkeley: University of California Press, 1983.

Heresbach, Conrad. *Foure Bookes of Husbandry.* Trans. Barnabe Goodge. London, 1577.

Herrick, Robert. "The Hock-Cart, or Harvest Home: To the Right Honorable Mildmay, Earl of Westmorland." In *Ben Johnson and the Cavalier Poets.* Ed. Hugh Maclean. New York: Norton, 1974, pp. 123–125.

Hibbard, G. R. "The Country House Poem of the Seventeenth Century." *Journal of the Warburg and Courtauld Institutes* 19: (1956): 159–174.

Higgins, Lynn A. and Brenda Silver, ed. *Rape and Representation.* New York: Columbia University Press, 1991.

Hill, Christopher. *The World Turned Upside Down: Radical Ideas During the English Revolution.* London and New York: Penguin, 1972.

Hogrefe, Pearl. *Tudor Women: Commoners and Queens.* Ames, Iowa: Iowa State Press, 1975.

Holmes, Martin. *Elizabethan London.* New York: Frederick A. Praeger, 1969.

Howard, Jean E. "Scripts and/versus Playhouses: Ideological Production and the Renaissance Public Stage." *Renaissance Drama* n.s. 20: (1989).

Howard, Jean E. *The Stage and Social Struggle in Early Modern England.* London and New York: Routledge, 1994.

Howard, Jean E. "Women as Spectators, Spectacles, and Paying Customers." In *Staging the Renaissance: Reinterpretations of Elizabethan and Jacobean Drama.* Ed. David Scott Kastan and Peter Stallybrass. New York and London: Routledge, 1991, pp. 68–74.

Howgego, James. *Printed Maps of London, circa 1553–1850.* Second ed. Folkestone, England: Wm. Dawson & Sons, 1964.

Hulme, Peter, and Neil Whitehead, eds. *Wild Majesty: Encounters with Caribs from Columbus to the Present Day: An Anthology.* Oxford: Clarendon Press, 1992.

Hyde, Ralph. *Gilded Scenes and Shining Prospects: Panoramic Views of British Towns 1575–1900.* New Haven: Yale Center for British Art, 1985.

Irigaray, Luce. *Speculum of the Other Woman.* Trans. Gillian C. Gill. Ithaca: Cornell University Press, 1985.

Irigaray, Luce. *This Sex Which Is Not One.* Trans. Catherine Porter with Carolyn Burke. Ithaca: Cornell University Press, 1985.

Jameson, Fredric. "Of Islands and Trenches: Naturalization and the Production of Utopian Discourse." *Diacritics* June (1977): 2–21.

Jed, Stephanie H. *Chaste Thinking: The Rape of Lucretia and the Birth of Humanism.* Bloomington and Indianapolis: Indiana University Press, 1989.

Johnston, George Burke. "Poems by William Camden: With Notes and Translations from the Latin." *Studies in Philology* 72:5 (1975): 86–105 and 134–143.

Jonson, Ben. *The Alchemist. The Complete Works of Ben Jonson.* Ed. G. A. Wilkes. Oxford: Clarendon, 1982.

Jonson, Ben. *Bartholomew Fair. The Complete Plays of Ben Jonson.* Ed. G. A. Wilkes. Oxford: Clarendon Press, 1982.

Jonson, Ben. *Ben Jonson (Complete Works).* 11 Vols. Ed. C. H. Herford and Percy Simpson. Oxford: Clarendon Press, 1925–1952.

Jonson, Ben. *The Staple of News,* 1626.

Jonson, Ben, George Chapman, and John Marston. *Eastward Ho!* Ed. C. G. Petter. New York: Norton, 1994.

Judges, Arthur Valentine, ed. *The Elizabethan Underworld; A Collection of Tudor and early Stuart tracts and ballads telling of the lives and misdoings of vagabonds, thieves, rogues, and cozeners, and giving some account of the operation of the criminal law.* New York: Routledge, 1930.

Kain, Roger J. P., and Elizabeth Baigent. *The Cadastral Map in the Service of the State: A History of Property Mapping.* Chicago and London: University of Chicago Press, 1992.

Kelsall, Malcolm. *The Great Good Place: The Country House and English Literature.* New York: Columbia University Press, 1993.

Kendrick, Christopher. "Agons of the Manor: 'Upon Appleton House' and Agrarian Capitalism." In *The Production of Renaissance Culture*. Ed. David Lee Miller, Sharon O'Dair, and Harold Weber. Ithaca and London: Cornell University Press, 1994, pp. 13–55.

Kingsford, Charles Letherbridge. "Essex House, formerly Leicester House and Exeter Inn." *Archaeology* 73:1923 (1923): 1–54.

Kinney, Arthur F. "Scottish History, the Union of the Crowns and the Issue of Right Rule: The Case of Shakespeare's *Macbeth*." In *Renaissance Culture in Context: Theory and Practice*. Ed. Jean R. Brink and William F. Gentrup. Aldershot, England: Scolar Press, 1993, pp. 18–53.

Knapp, Jeffrey. *An Empire Nowhere: England, America, and Literature from Utopia to The Tempest*. Berkeley: University of California Press, 1992.

Koch, Mark. "Ruling the World: The Cartographic Gaze in Elizabethan Accounts of the New World." *Early Modern Literary Studies* 4.2 / Special Issue 3 (September, 1998): 11.1–39 <http://www.shu.ac.uk/emls/04-2/kochruli.htm>.

Kolodny, Annette. *The Lay of the Land*. Chapel Hill: University of North Carolina Press, 1975.

Krontiris, Tina. *Oppositional Voices: Women as Writers and Translators of Literature in the English Renaissance*. London and New York: Routledge, 1992.

Lachmann, Richard. *From Manor to Market: Structural Change in England, 1536–1640*. Madison: University of Wisconsin Press, 1987.

Lambarde, William. *A Perambulation of Kent: Conteining the Description, Hystorie, and Customes of That Shire. Written in the Yeere 1570*. London: Baldwin, Cradock, and Joy, (1576) 1826.

Lanyer, Aemelia. *Salve Deus Rex Judaeorum*. 6th ed., 1611.

Lasdun, Susan. *The English Park: Royal, Private & Public*. New York: Vendome Press, 1992.

Lefebvre, Henri. *The Production of Space*. Trans. Donald Nicholson-Smith. Oxford and Cambridge, Mass.: Blackwell, 1991.

Leggatt, Alexander. *Ben Jonson: His Vision and His Art*. London and New York: Methuen, 1981.

Lehner, Ernst and Johanna. *How They Saw the New World*. Ed. Gerard L. Alexander. New York: Tudor Publishing Company, 1966.

Leigh, Valentine. *The Moste Profitable and commendable Science, of Surueiying of Landes, Tenementes, and Hereditamentes: drawen and collected by the industrie of Valentine Leigh*. London: Andrew Maunsell, 1578.

Leinwand, Theodore B. *The City Staged: Jacobean Comedy, 1603–1613*. Madison: University of Wisconsin Press, 1986.

Leinwand, Theodore B. "London Triumphing: The Jacobean Lord Mayor's Show." *Clio* 11:2 (1982): 137–153.

Leland, John. *Cygnea Cantio*. In *The Itinerary of John Leland the Antiquary*. Ed. Thomas Hearne. 9 Vols. Oxford, (1545) 1770. Vol. 9. pp. 1–108.

Leland, John. *The Itinerary of John Leland in or about the years 1535–1543*. 5 Vols. Ed. Lucy Toulmin Smith. Carbondale, Ill.: Southern Illinois University Press, 1964.

Lenman, Bruce P. "England, the International Gem Trade and the Growth of Geographical Knowledge from Columbus to James I." In *Renaissance Culture in Context: Theory and Practice.* Ed. Jean R. Brink and William F. Gentrup. Cambridge: Cambridge University Press, 1993, pp. 86–99.

Lestringant, Frank. "Geneva and America in the Renaissance: The Dream of the Huguenot Refuge 1555–1600." *Sixteenth Century Journal* XXVI:2. Summer (1995): 285–295.

Lestringant, Frank. *Mapping the Renaissance World: The Geographical Imagination in the Age of Discovery.* Trans. David Fausett. Berkeley: University of California Press, 1994.

Levin, Richard, ed. *Michalemas Term.* Lincoln: University of Nebraska Press, 1966.

Levine, Laura. *Men in Women's Clothing: Anti-theatricality and Effeminization, 1579–1642.* Cambridge: Cambridge University Press, 1994.

Lewalski, Barbara. "Re-writing Patriarchy and Patronage: Margaret Clifford, Anne Clifford, and Aemilia Lanyer." *Yearbook in English Studies* 21: (1991): 87–106.

Lindley, David, ed. *Court Masques: Jacobean and Caroline Entertainments, 1605–1640.* Oxford and New York: Oxford University Press, 1995.

Lister, Raymond. *How to Identify Old Maps and Globes.* Hamden, Connecticut: Archon Books, 1965.

Littleton, Thomas. *Littletons Tenures in English.* London, 1586.

Lloyd, Robert. "A Look at Images." *Annals of the Association of American Geographers* 72:4 (1982): 532–48.

Lobanov-Rostovsky, Sergei. "*The Triumphes of Golde:* Economic Authority in the Jacobean Lord Mayor's Show." *ELH* 60: (1993): 879–898.

London, Corporation of. *Collage.* <http://collage.nhil.com>.

Loomis, Albert J., S. J. "An Armada Pilot's Survey of the English Coastline, October 1597." *The Mariner's Mirror* 49:1 Feb (1963): 288–300.

Machiavelli, Niccolo. *The Art of War.* Trans. Ellis Farneworth. New York: Bobbs-Merrill, 1965.

Machiavelli, Niccolo. *The Discourses.* Ed. Bernard Crick. Trans. Leslie J. Walker, S.J. Revised, 1987 ed. Hammondsworth, England: Penguin, 1970.

Machiavelli, Niccolo. *The Prince.* Trans. Luigi Ricci, revised by E. R. P. Vincent. 1532 ed. New York: New American Library, 1980.

Malcolmson, Christina. "The Garden Enclosed / The Woman Enclosed: Marvell and the Cavalier Poets." In *Enclosure Acts: Sexuality, Property, and Culture in Early Modern England.* Ed. Richard Burt and John Michael Archer. Ithaca: Cornell University Press, 1994, pp. 251–269.

Manley, Lawrence. "From Matron to Monster: Tudor-Stuart London and the Languages of Urban Description." In *The Historical Renaissance: New Essays on Tudor and Stuart Literature and Culture.* Ed. Heather Dubrow and Richard Strier. Chicago and London: University of Chicago Press, 1988, pp. 347–374.

Manley, Lawrence. *Literature and Culture in Early Modern London.* Cambridge: Cambridge University Press, 1995.

Manley, Lawrence, ed. *London in the Age of Shakespeare: An Anthology*. University Park: University of Pennsylvania Press, 1986.

Manley, Lawrence. "Of Sites and Rites." In *The Theatrical City: Culture, Theatre, and Politics in London, 1576–1649*. Ed. David L. Smith, Richard Strier, and David Bevington. Cambridge and New York: Cambridge University Press, 1995, pp. 35–54.

Mantilla, Luis Carlos. *Una Expressión Artística Inspirada en Historias Primigenias de América: Catorce Grabados Escogidos de la "Idea Verdadera y Genuina [. . .]" por Teodoro de Bry (Edicion Facsimilar Sobre la Impresión de 1602*. Bogotá: Instituto Caro y Cuervo, 1988.

MapHist, the Map History Discussion List. <http://www.maphist.nl>.

Marchitello, Howard. "Political Maps: The Production of Cartography and Chorography in Early Modern England." In *Cultural Artifacts and the Production of Meaning: The Page, The Image, and The Body*. Ed. Margaret J. M. Ezell and Katherine O'Brien O'Keeffee. Ann Arbor: University of Michigan Press, 1994, pp. 13–40.

Marin, Louis. *Utopics: Spatial Play*. Trans. Robert A. Vollrath. Atlantic Highlands, N.J.: Humanities Press, 1984.

Marks, Stephen Powys. *The Map of Mid Sixteenth Century London: An Investigation into the Relationship Between a Copper-engraved Map and its Derivatives*. London: London Topographical Society, 1964.

Marlier, Georges. *Pierre Brueghel Le Jeune*. Bruxelles: Robert Finck, 1969.

Marotti, Arthur F. *Manuscript, Print, and the English Renaissance Lyric*. Ithaca and London: Cornell University Press, 1995.

Marvell, Andrew. *The Complete Poems*. Ed. Elizabeth Story Donno. London: Penguin, 1972.

McClung, William A. *The Country House in English Renaissance Poetry*. Berkeley: University of California Press, 1977.

McCoy, Richard. *The Rites of Knighthood: The Literature and Politics of Elizabethan Chivalry*. Berkeley: University of California Press, 1989.

McGrath, Lynette. "'Let Us Have Our Libertie Againe': Amelia Lanier's 17th-Century Feminist Voice." *Women's Studies* 20: (1992): 331–348.

McRae, Andrew. *God Speed the Plough: The Representation of Agrarian England 1500–1660*. Cambridge: Cambridge University Press, 1996.

McRae, Andrew. "'On the Famous Voyage': Ben Jonson and Civic Space." *Early Modern Literary Studies* 4.2 / Special Issue 3 (1998): 8.1–31 <http://www.shu.ac.uk/emls/04-2/mcraonth.htm>.

Mercator, Gerard. *Atlas sive Cosmographicae meditationes de fabrica mundi et fabricati figura*. Dusseldorf: A. Busius, 1595.

Mercator, Gerard. *World Map*, 1569.

Merleau-Ponty, Marcel. *Phenomenology of Perception*. Trans. Colin Smith. London and New York: Routledge, 1995.

Middleton, Thomas. *A Chaste Maid in Cheapside*, 1613.

Middleton, Thomas. *Michaelmas Term*. Ed. Richard Levin. Lincoln: University of Nebraska Press, 1966.

Middleton, Thomas. *The Works of Thomas Middleton.* 8 Vols. Ed. A. H. Bullen. London: John C. Nimmo, 1895.

Middleton, Thomas, and Thomas Dekker. *The Roaring Girl.* Ed. Paul A. Mulholland. Manchester: Manchester University Press, 1987.

Mikalachki, Jodi. *The Legend of Boadicea: Gender and Nation in Early Modern England.* London and New York: Routledge, 1998.

Milton, John. *Complete Poems and Major Prose.* Ed. Merritt Y. Hughes. New York: Bobbs-Merrill, 1957.

Moi, Toril, ed. *The Kristeva Reader.* New York: Columbia University Press, 1986.

Monmonier, Mark. *Drawing the Line: Tales of Maps and Cartocontroversy.* New York: Henry Holt, 1995.

Montrose, Louis Adrian. "The Work of Gender in the Discourse of Discovery." *Representations* 33: Winter (1991): 1–41.

Montrose, Louis Adrian. "The Elizabethan Subject and the Spenserian Text." In *Literary Theory/Renaissance Texts.* Ed. Patricia Parker and David Quint. Baltimore: Johns Hopkins University Press, 1986, pp. 301–240.

Morden, Robert. *Morden's Playing Cards.* Facsimile of 1676 ed. Lympne Castle, Kent: Harry Margary, 1972.

Morden, Robert. *Playing Cards.* London: British Library, Print Library, 1676.

More, Thomas. *Utopia.* New Haven: Yale University Press, 1964.

Morgan, Victor. "Lasting Image of the Elizabethan Era." *The Geographical Magazine* 52 March (1980): 401–8.

Morley, Henry, ed. *Ideal Commonwealths: Comprising: More's* Utopia, *Bacon's* New Atlantis, *Campanella's* City of the Sun *and Harrington's* Oceana. Sawtry, Cambridgeshire: Dedalus, 1988.

Morris, Mary, ed. *Maiden Voyages: Writings of Women Travelers.* New York: Vintage, 1993.

Mueller, Janel. "The Feminist Poetics of Aemilia Lanyer's *Salve Deus Rex Judaeorum.*" In *Feminist Measures: Soundings in Poetry and Theory.* Ed. Lynne Keller and Christanne Miller. Ann Arbor: University of Michigan Press, 1994, pp. 208–236.

Mullaney, Steven. *The Place of the Stage: License, Play, and Power in Renaissance England.* Chicago and London: University of Chicago Press, 1988.

Neale, J. E. "Introduction." In *The Quenes Maiesties Passage through the Citie of London to Westminster the Day before her Coronacion.* Ed. James M. Osborn. New Haven: Yale University Press, 1960, pp. 7–15.

Newey, Vincent, and Ann Thompson, eds. *Literature and Nationalism.* Savage, Md.: Barnes & Noble, 1991.

Nichols, John, ed. *The Progresses and Public Processions of Queen Elizabeth.* London: John Nichols, 1823.

Nicolson, Nigel, ed. *The Counties of Britain: A Tudor Atlas by John Speed.* London: Pavilion Books, 1988.

Nora, Pierre. "Between Memory and History: *Les Lieux de Memoire.*" *Representations* 26 Spring (1989): 7–25.

Nora, Pierre. *Les Lieux de Memoire.* 3 Vols. Paris: Gallimard, 1984.

Norden, John. *Chorographical Description of Several Shires,* ca. 1595. British Library, *Add. MS 31853.*

Norden, John. *Chorographicall discription of Middlesex, Essex, Surrey, Sussex, Hampshire and the Channel Islands,* 1595.

Norden, John. *Nordens Preparatiue to his Speculum Britaniae. Intended A reconciliation of sundrie propositions by diuers persons tendred, concerning the same.* London, 1596.

Norden, John. *Speculi Britaniae, Pars: The Description of Hartfordshire.* 1598.

Norden, John. *Speculum Britanniae, The first parte. An historicall, & chorographicall disciption of Middlesex, Wherein are also alphabeticallie sett downe the names of the cyties, townes, parishes, hamletes, howses of name & c. W.th direction spedelie to finde anie place desired in the mappe & the distance betweene place and place without compasses.* 1593.

Norden, John. *Surveiors Dialogue, Very profitable for all men to peruse, but especially for Gentlemen, Farmers, and Husbandmen, that shall either haue occasion, or be willing to buy, hire, or sell Lands: As in the ready and perfect Surueying of them, with the manner and Method of keeping a Court of Survey with many necessary rules, and familiar Tables to that purpose.* 3rd ed. London: Thomas Snodham, 1618.

Nosworthy, J. M., ed. *Cymbeline.* New York: Routledge, 1991.

Olds, Sharon. "Topography." *The Gold Cell.* New York: Knopf, 1987, p. 58.

Orgel, Stephen. *The Illusion of Power: Political Theater in the English Renaissance.* Berkeley: University of California Press, 1975.

Oruch, Jack B. "Spenser, Camden, and the Poetic Marriage of Rivers." *Studies in Philology* 64: (1967): 606–24.

Osborn, James M., ed. *The Quenes Maiesties Passage through the Citie of London to Westminster the Day before her Coronacion, Anno 1558.* New Haven: Yale University Press, 1960.

Osgood, Charles Grosvenor. "Spenser's English Rivers." *Transactions of the Connecticut Academy of Arts and Sciences* 23 (1920): 65–108.

Paglia, Camille. *Sexual Personae: Art and Decadence from Nefertiti to Emily Dickinson.* New York: Vintage, 1990.

Palsky, Gilles. "Borges, Carroll, et la Cart au 1/1." *Cybergeo* 106 (1999): <http://www.cybergeo.presse.fr/cartogrf/texte1/jborges.htm>.

Parker, Andrew, et al., eds. *Nationalisms and Sexualities.* New York and London: Routledge, 1992.

Parker, M. Pauline. *The Allegory of* The Faerie Queene. Oxford: Clarendon Press, 1960.

Parker, Patrica. *Literary Fat Ladies: Rhetoric, Gender, Property.* New York: Methuen, 1987.

Parker, Patricia. "Shakespeare and rhetoric: 'dilation' and 'delation' in *Othello*." In *Shakespeare and the Question of Theory*. Ed. Patricia Parker and Geoffrey Hartman. New York: Routledge, 1990, pp. 54–74.

Partridge, Eric. *Shakespeare's Bawdy*. 3rd ed. London and New York: Routledge, 1990.

Paster, Gail Kern. *The Body Embarrassed: Drama and the Disciplines of Shame in Early Modern England*. Ithaca: Cornell University Press, 1993.

Paster, Gail Kern. *The Idea of the City in the Age of Shakespeare*. Athens: University of Georgia Press, 1985.

Petter, C. G., ed. *Eastward Ho!* London: A & C Black, 1994.

Porter, Thomas. *The Newest and Exactest MAPP of the most Famous Citties LONDON and WESTMINSTER with their Suburbs; and the manner of their streets: With the Names of the Chiefest of them Written at Length and Numbers set in the rest in sted of Names. The which Names are at Length in the Table with Numbers how to Guide them Readily So that it is a ready Helpe or Guide to direct Country-men and strangers to finde the nearest way from one place to another by T. Porter*, 1655.

Portinaro, Pierluigi, and Franco Knirsch. *The Cartography of North America: 1500–1800*. New York: Crown Publishers, 1987.

Prockter, Adrian, and Robert Taylor. *The A to Z of Elizabethan London*. London: London Topographical Society, 1979.

Ptolemy, Claudius. *The Geography*. Trans. and editor Edward Luther Stevenson. Based on 1460 ed. New York: Dover, 1991.

Quilligan, Maureen. "Feminine Endings: The Sexual Politics of Sidney's and Spenser's Rhyming." In *The Renaissance Englishwoman in Print: Counterbalancing the Canon*. Ed. Anne M. Haselkorn and Betty S. Travitsky. Amherst: University of Massachusetts Press, 1990, pp. 311–326.

Quinlan-McGrath, Mary. "Caprarola's Sala della Cosmografia." *Renaissance Quarterly* 50:4 (1997): 1045–1100.

Quint, David. *Origin and Originality in Renaissance Literature*. New Haven and London: Yale University Press, 1983.

Rabasa, José. *Inventing America: Spanish Historiography and the Formations of Eurocentrism*. Norman and London: University of Oklahoma Press, 1993.

Ralegh, Sir Walter. *The Discoverie of the Large, Rich, and Bewtiful Empyre of Guiana, with a relation of the great and Golden Citie of Manoa (which the Spanyards call El Dorado) And the provinces of Emeria, Arromaia, Amapaia, and other Countries, with their rivers, adjoyning*. facs. 1596 ed. Amsterdam: Voorurgwal, 1968.

Ralegh, Sir Walter. "The Discovery of the Large, Rich and Beautiful Empire of Guiana, With a relation of the Great and Golden City of Manoa (which the Spaniards call El Dorado) And the provinces of Emeria, Arromaia, Amapaia and other Countries, with their river, adjoining." In *Sir Walter Ralegh: Selected Writings*. Ed. Gerald Hammond. Great Britain: Carcanet Press, 1984, pp. 76–123.

Ralegh, Sir Walter. *The Poems of Sir Walter Ralegh*. Ed. Agnes M. C. Latham. Cambridge: Harvard University Press, 1951.

Rathborne, Aaron. *The Surveyor in Four Bookes*. London: W. Burre, 1616.

Ravenhill, William, ed. *Christopher Saxton's 16th Century Maps: The Counties of England and Wales*. Shrewsbury: Chatsworth Library, 1992.

Retamar, Roberto Fernandez. *Caliban and Other Essays*. Trans. Edward Baker. Minneapolis: University of Minnesota Press, 1989.

Ringler, William A., Jr. *The Poems of Sir Philip Sidney*. Oxford: Clarendon Press, 1962.

Roberts, Jeanne Addison. *The Shakespearean Wild: Geography, Genus, and Gender*. Lincoln: University of Nebraska Press, 1991.

Roche, Thomas P., Jr. *The Kindly Flame: A Study of the Third and Fourth Books of Spenser's* Faerie Queene. Princeton, New Jersey: Princeton University Press, 1964.

Rogers, John. "The Enclosure of Virginity: The Poetics of Sexual Abstinence in the English Revolution." In *Enclosure Acts: Sexuality, Property, and Culture in Early Modern England*. Ed. Richard Burt and John Michael Archer. Ithaca and London: Cornell University Press, 1994, pp. 229–250.

Royal Museum of Fine Arts, Antwerp. *America, Bride of the Sun*. Brussels: Imschoot Books, 1992.

Rubenstein, Frankie. *A Dictionary of Shakespeare's Sexual Puns and their Significance*. 2nd ed. New York: St. Martin's Press, 1995.

Rye, William Benchley, ed. *England as Seen by Foreigners*. London, 1865.

Sackville-West, Vita. *The Diary of Anne Clifford*. London: William Heinemann Ltd., 1923.

Sanford, Rhonda Lemke. "A Room Not One's Own: Feminine Geography in Cymbeline." In *Playing the Globe: Genre and Geography in English Renaissance Drama*. Ed. John Gillies and Virginia Mason Vaughan. Madison and Teaneck, N.J. and London: Fairleigh Dickinson University Press and Associated University Presses, 1998, pp. 63–85.

Saunders, J. W. "The Stigma of Print: A Note on the Social Bases of Tudor Poetry." *Essays in Criticism* 1: (1951): 139–164.

Saxton, Christopher. *The Counties of England and Wales*, 1579.

Schnell, Lisa. "'So Great a Difference Is There in Degree': Aemilia Lanyer and the Aims of Feminist Criticism." *Modern Language Quarterly* 57:1 (1996): 23–35.

Schulz, Juergen. "Maps as Metaphors: Mural Map Cycles of the Italian Renaissance." In *Art and Cartography*. Ed. David Woodward. Chicago and London: University of Chicago Press, 1978, pp. 97–122.

Sedgwick, Eve Kosofsky. *Between Men: English Literature and Male Homosocial Desire*. New York: Columbia University Press, 1985.

Shakespeare, William. *Antony and Cleopatra*. Ed. M. R. Ridley. London: Routledge, 1954.

Shakespeare, William. *Cymbeline*. Ed. J. M. Nosworthy. London and New York: Routledge, 1991.

Shakespeare, William. *The First Part of King Henry IV.* Ed. A. R. Humphreys. New York and London: Routledge, 1991.

Shakespeare, William. *Richard II.* Ed. Peter Ure. London: Routledge, 1991.

Shakespeare, William. *The Riverside Shakespeare.* Ed. G. Blakemore Evans. 2nd ed. Boston and New York: Houghton Mifflin Company, 1997.

Shakespeare, William. *The Second Part of King Henry IV.* Ed. A. R. Humphries. New York and London: Routledge, 1991.

Shakespeare, William. *The Taming of the Shrew.* Ed. Brian Morris. London: Methuen, 1981.

Shakespeare, William. *The Tempest.* Ed. Frank Kermode. London: Methuen, 1958.

Shakespeare, William. *Troilus and Cressida.* Ed. Kenneth Palmer. London: Methuen, 1982.

Shirley, Rodney W. *Early Printed Maps of the British Isles, A Bibliography 1477–1650.* King of Prussia, Penn.: W. Graham Arader III, 1980.

Sidney, Philip. *An Apology for Poetry.* Ed. Forrest G. Robinson. 1970 ed. Indianapolis: Bobbs-Merrill, 1595.

Sidney, Philip. *The Four Foster Children of Desire.* In *Entertainments for Elizabeth I.* Ed. Jean Wilson. Totawa, N.J.: Brewer, Rowman, and Littlefirld, 1980.

Sidney, Sir Philip. *The Lady of May.* In *Miscellaneous Prose of Sir Philip Sidney.* Ed. Katherine Duncan-Jones and Jan Van Dorsten. Oxford: Clarendon, 1973, pp.

Sidney, Philip. *Miscellaneous Prose of Sir Philip Sidney.* Eds. Katherine Duncan-Jones and Jan van Dorsten. Oxford: Clarendon Press, 1973.

Siebers, Tobin, ed. *Heterotopia: Postmodern Utopia and the Body Politic.* Ann Arbor: University of Michigan Press, 1994.

Silberman, Lauren. *Transforming Desire: Erotic Knowledge in Books III and IV of The Faerie Queene.* Berkeley: University of California Press, 1995.

Slights, William W. E. *Ben Jonson and the Art of Secrecy.* Toronto, Buffalo, London: University of Toronto Press, 1994.

Smith, Lucy Toulmin, ed. *The Itinerary of John Leland in or about the Years 1535–1543.* 5 Vols. Carbondale: Southern Illinois University Press, 1964.

Smith, William. *[Twelve County Maps].* London: Hans Woutneel, 1602–3.

Speed, John. *The Counties of Britain.* 1616 ed. London, 1611–1612.

Speed, John. *The Theatre of the Empire of Great Britaine,* 1611.

Spenser, Edmund. *The Faerie Queene.* Ed. A. C. Hamilton. London and New York: Longman, 1977.

Spenser, Edmund. "Three proper wittie familiar Letters, lately passed betwene two vniuersitie men, touching the Earthquake in April last, and our English reformed Versifying, [Letter III]." In *The Works of Edmund Spenser: A Variorum Edition.* Ed. Charles Grovenor Osgood, Edwin Greenlaw, Frederick Padelford, Rudolf Gottfried (special editor, Vol. 9). Baltimore: Johns Hopkins Press, 1949. Vol. 9, pp. 15–17.

Spenser, Edmund. *A View of the Present State of Ireland*. Ed. W. L. Renwick. Oxford: Clarendon Press, 1970.

Spenser, Edmund. *A View of the State of Ireland*. Ed. Andrew Hadfield and Willy Maley. Oxford: Blackwell, 1997.

Spenser, Edmund. *The Yale Edition of the Shorter Poems of Edmund Spenser.* Eds. William A. Oram, et al. New Haven and London: Yale University Press, 1989.

Spivak, Gayatri Chakravorty. "Woman in Difference: Mahasweta Devi's 'Douloti the Bountiful.'" *Cultural Critique,* Winter (1992): 105–128.

Spring, Eileen. *Law, Land & Family: Aristocratic Inheritance in England, 1300 to 1800*. Chapel Hill: University of North Carolina Press, 1993.

Stallybrass, Peter. "Patriarchal Territories: The Body Enclosed." In *Rewriting the Renaissance: The Discourses of Sexual Difference in Early Modern Europe*. Ed. Margaret W. Ferguson, Maureen Quilligan, and Nancy Vickers. Chicago: University of Chicago Press, 1986, pp. 123–142.

Stallybrass, Peter and Allon White. *The Politics and Poetics of Transgression*. Ithaca: Cornell University Press, 1986.

Steele, Ian K. *Warpaths: Invasions of North America*. New York and Oxford: Oxford University Press, 1994.

Stevenson, Jane, and Peter Davidson, eds. *Early Modern Women Poets: An Anthology*. Oxford and New York: Oxford University Press, 2001.

Stewart, George. *The Technique of English Verse*. Port Washington, N.Y.: Kennikat Press, 1966.

Stimpson, Catharine R. "Shakespeare and the Soil of Rape." In *The Woman's Part: Feminist Criticism of Shakespeare*. Ed. Carolyn Ruth Swift Lenz, Gayle Greene, and Carol Thomas Neely. Urbana and Chicago: University of Illinois Press, 1983, pp. 56–64.

Stow, John. *A Survey of London Written in the Year 1598*. Ed. Henry Morley. Phoenix Mill: Alan Sutton, 1994.

Strong, Roy. *The Cult of Elizabeth: Elizabethan Portraiture and Pageantry*. Berkeley and Los Angeles: University of California Press, 1977.

Strong, Roy. *Portraits of Queen Elizabeth I*. Oxford: Clarendon Press, 1964.

Strong, Roy. *Royal Gardens*. New York: Pocket Books, 1992.

Stubbes, Phillip. *The Anatomie of Abuses*. Facsimile of 1583 ed. New York and London: Garland, 1973.

Sullivan, Garrett A. Jr. *The Drama of Landscape: Land, Property, and Social Relations on the Early Modern Stage*. Stanford: Stanford University Press, 1998.

Sullivan, Garrett A. Jr. "Civilizing Wales: *Cymbeline*, Roads and the Landscapes of Early Modern Britain." *Early Modern Literary Studies* 4:2 / Special Issue 3 (1998): <http://www.humanities.ualberta.ca/emls/04–2/sullshak.htm>.

Todorov, Tzvetan. *The Conquest of America: The Question of the Other.* Trans. Richard Howard. 1992 ed. New York: Harper Perennial, 1982.

Traub, Valerie. "Mapping the Global Body." Paper Presented at the Conference: Paper Landscapes: Maps, Texts and The Construction of Space, 1500–1700. Queen Mary and Westfield College, London, England, 1997.

Travitsky, Betty. "The 'Wyll and Testament' of Isabella Whitney." *English Literary Renaissance* 10:Winter (1980): 76–94.

Turnbull, David. "Cartography and Science in Early Modern Europe: Mapping the Construction of Knowledge Spaces." *Imago Mundi* 48: (1996): 5–24.

Turnbull, David. *Maps are Territories: Science is an Atlas: A portfolio of exhibits.* 1993 ed. Chicago: University of Chicago Press, 1989.

Turner, James. *The Politics of Landscape: Rural Scenery and Society in English Poetry 1630–1660.* Cambridge: Harvard University Press, 1979.

Tyacke, Sarah, ed. *English Map-making 1500–1650.* London: The British Library, 1983.

Unknown. *International Playing Cards.* London: British Museum (Schrieber Collection: O'Donoghue E. 44), ca. 1676.

Vallans, William. *A tale of tvvo swannes. VVherein is comprehended the original and increase of the riuer Lee commonly called Ware-riuer: together, with the antiquitie of sundrie places and townes seated vpon the same. Pleasant to be read, and not altogether vnprofitable to bee vnderstood.* London: Roger Ward, 1590.

Vickers, Nancy. ""The Blazon of Sweet Beauty's Best": Shakespeare's *Lucrece.*" In *Shakespeare and the Question of Theory.* Ed. Patricia Parker and Geoffrey Hartman. New York: Routledge, 1985, pp. 95–115.

Voekel, Swen. "'Upon the Suddaine View': State, Civil Society and Surveillance in Early Modern England." *Early Modern Literary Studies* 4.2 / Special Issue 3 (1998): 2.1–27 <http://www.shu.ac.uk/emls/04–2/voekupon.htm>.

W.B. *[A pack of playing-cards with maps of the counties of England and Wales, signed W.B. inuent. 1590],* British Museum, Print Room, Schrieber Collection.

Wall, Wendy. *The Imprint of Gender: Authorship and Publication in the English Renaissance.* Ithaca: Cornell University Press, 1993.

Wall, Wendy. "Isabella Whitney and the Female Legacy." *ELH* 58: (1991): 35–62.

Waller, Gary. "Struggling into Discourse: The Emergence of Renaissance Women's Writing." In *Silent but for the Word.* Ed. Margaret Hannay. Kent, Ohio: Kent State University Press, 1985, pp. 238–256.

Wallerstein, Immanuel. *The Modern World-System: Capitalist Agriculture and the Origins of the European World-Economy in the Sixteenth Century.* New York & London: Academic Press, 1974.

Wallerstein, Immanuel. *The Modern World-System II: Mercantilism and the Consolidation of the European World-Economy, 1600–1750.* New York: Academic Press, 1980.

Wands, John Millar. *Another World and Yet the Same: Bishop Joseph Hall's Mundus Alter et Idem.* New Haven: Yale University Press, 1991.

Wayne, Don E. *Penshurst: The Semiotics of Place and the Poetics of History.* Madison: University of Wisconsin Press, 1984.

Weimann, Robert. *Author's Pen and Actor's Voice: Playing and Writing in Shakespeare's Theatre.* Cambridge: Cambridge University Press, 2000.

Weimann, Robert. *Shakespeare and the Popular Tradition in the Theater: Studies in the Social Dimension of Dramatic Form and Function*. Ed. Robert Schwartz. Baltimore and London: Johns Hopkins University Press, 1978.

Whitfield, Peter. *The Image of the World: 20 Centuries of World Maps*. London: The British Library, 1994.

Whitney, Isabella. "'The Author . . . Maketh Her Will and Testament,' including 'A Communication Which the Author had to London Before She Made Her Will' and 'The Manner of Her Will and What She Left to London and to All Those in It at Her Departing'." In *A sweer nosegay or pleasnt posye. Contayning a hundred and ten Phylosophicall flowers*. London: 1573.

Wickham, Glynne. *Early English Stages 1300–1600*. 2 Vols. London: Routledge & Kegan Paul, 1963.

Wilford, John Noble. *The Mapmakers: The Story of the Great Pioneers in Cartography from Antiquity to the Space Age*. New York: Vintage Books, 1981.

Williams, Penry. "The Crown and the Counties." In *The Reign of Elizabeth I*. Ed. Christopher Haigh. Athens: University of Georgia Press, 1985.

Williams, Raymond. *The Country and the City*. New York: Oxford University Press, 1973.

Wood, Denis. *The Power of Maps*. New York: Guilford Publications, 1992.

Woodbridge, Linda. "Palisading the Elizabethan Body Politic." *Texas Studies in Literature and Language* 33:3 (1991): 327–354.

Woods, Susanne. "Aemilia Lanyer and Ben Jonson: Patronage, Authority, and Gender." *The Ben Jonson Journal: Literary Contexts in the Age of Elizabeth, James, and Charles* 1: (1994): 15–30.

Woods, Susanne, ed. *The Poems of Aemilia Lanyer: Salve Deus Rex Judaeorum*. New York and Oxford: Oxford University Press, 1993.

Woodward, David. *Art and Cartography: Six Historical Essays*. Chicago and London: University of Chicago Press, 1987.

Woolway, Joanne. "Spenser and the Culture of Place." *Interactive Early Modern Literary Studies* (1996): <http://www.shu.ac.uk./schools/cs/emls/iemls/conf/texts/woolway.html>.

Yates, Frances A. *The Art of Memory*. Chicago: University of Chicago Press, 1966.

Yates, Frances A. *Theatre of the World*. London and New York: Routledge & Kegan Paul, 1987.

Yates, Julian. "The Geometry of Forgetting: Maps, Mapping and the Culture of Print in Early Modern England." Paper Presented at the Conference: Paper Landscapes: Maps, Texts and the Construction of Space 1500–1700. Queen Mary and Westfield College, University of London, 1997.

Zamora, Margarita. *Reading Columbus*. Ed. Roberto Gonzalez Echevarría. Berkeley: University of California Press, 1993.

Ziegler, Georgianna. "My Lady's Chamber: Female Space, Female Chastity in Shakespeare." *Textual Practice* 4(1) Spring (1990): 73–90.

INDEX